FTC ROBOTICS:

TIPS, TRICKS, STRATEGIES, AND SECRETS

2013-14 EDITION

FTC Robotics:

Tips, Tricks, Strategies, and Secrets

2013-14 Edition

The Pope John XXIII High School Robotics Team

Editor-in-Chief: Caitlyn Cherepakhov

Cover Illustration: Jacob Scordato

Interior Illustrations: Joseph Vengen, Jacquelyn Elio

Index: Sonia Geba

Please send comments or your own tips, tricks,
strategies, and secrets for future versions of this book to:

pj.ftc.tips@gmail.com

If we use your material you will receive
attribution in the book as well as free copies.

If any of the tips in this book conflict with manufacturer instructions for tools, materials, robot kit components, or other items then always follow the manufacturer's instructions. Always use common sense and standard safety precautions. Adult supervision is required for robot construction activities.

ISBN: 1492763950 REV: E3v4a 10-18-2013

ISBN-13: 978-1492763956

Praise from FTC Coaches for *FTC Robotics: Tips, Tricks Strategies, and Secrets*:

"As an FRC mentor, and new FTC coach, this book represents my best find of the year. Well organized and easy to digest, it provides beginner to expert tips in an easy to read and refer back to format. There is no other similar resource."

—Max G.

"As a rookie coach I learned so much from the book. Thanks to you and your team members and other contributors for taking time to share tips and strategy ideas with others. I am asking every team member to read the book . . . You made our FTC experience more enjoyable and rewarding."

—Sastry D., City, New Jersey

". . . I cannot tell you how helpful it has been to read this book. There are very, very many "tricks of the trade" given by veteran teams that really know what they're doing, having competed and won at top levels of FTC competition. The concrete examples given, the completely interesting & encouraging style, the sensible organization as well as the sheer volume and breadth of useful information you can pick up from reading this book all lend to its being a terrific resource for any FTC team--not just rookies."

—Doreen W., Auburn, California

"Our rookie team found this book to be an invaluable resource. It has enabled us to gain knowledge that would have taken us years to gather; it should be required reading for all new FTC teams!"

—Chad V., New Kensington, Pennsylvania

Table of Contents

Preface to the 2013-14 Edition

This extensively updated edition of *FTC Robotics: Tips, Tricks, Strategies, and Secrets* incorporates copious new material, including:

- Many new and updated illustrations and photographs
- Many new tips, including 3D printing and alternative building materials recently allowed in the competitions
- For the first time, an Index to help locate information quickly
- An updated section on plastic fabrication techniques, including thermoforming of new plastics allowed for the first time in the 2012/13 season.
- A new section on new building materials allowed in the 2012/13 and 2013/2014 seasons including extruded aluminum beams
- Updates based on our experiences with the 2012/13 game: "Ring it Up!"
- Expanded discussions, corrections and clarifications of many of the prior edition's tips

As with the first two editions, our primary focus is to provide practical advice for both novice and veteran teams on how to successfully compete in the FIRST Tech Challenge competition.

We were surprised at the rapid acceptance of the first two editions by teams from all over the world. In the first eight months after publication nearly a thousand copies went into print (including 200 we gave away to rookie teams). We received enquiries from all over the world and several major FTC affiliates informed us of their plans to distribute the book throughout their regions in 2011/12 season. The book has maintained an exceptionally high rating on Amazon.com, and several times has been ranked first in the Technology and Robotics categories there. Lego Education contacted our team and asked us to help brainstorm their new Tetrix kit materials. We even started to receive "fan mail" from across the country!

Perhaps the most satisfying of all these events occurred at the World Championships in May, 2011. Each day of the competition, teams would come up to our pit and thank us for producing the book. One team went so far as to say that without the book they would never have qualified for Worlds at all. Sometimes people would see our "Pope John Robotics" t-shirts while walking through the pits and gleefully shout, "I read your book!"

It's terrific to work hard on a project and see that it helped real people solve real problems; that it had a concrete impact.

Even though we may be off to a good start, our work has barely begun. This new edition is meant to raise the bar even further by providing information on topics we didn't even mention in the first edition. We have also included tips and suggestions we received from other teams and coaches, and we are very thankful for all the great teams we have talked to along the way.

We hope you find the new edition helpful in your journey through the FTC competitions. Good luck!

Introduction

FIRST Tech Challenge (FTC) is the fastest growing level of FIRST robotics, expanding from under 1,000 teams in the 2008/09 season to nearly 2,000 teams in the 2013/14 season. The FTC competition offers a level of technical sophistication nearly equal to the "big" robot league, FIRST Robotics Competition (FRC), but is more practical for many schools and organizations due to dramatically lower costs for materials and tournament fees, as well as the ability for teams to maintain a full practice field in a relatively small space (12 by 12 feet for FTC as opposed to nearly filling a high school gym for FRC).

This book began as a community outreach project by the Pope John XXIII Regional High School FTC Robotics program, Teams 247 and 248. The idea originated when the coach found that new FTC teams were hungry for information about how to be successful. He was frequently bombarded with questions ranging from "where did you get that rolling cart?" to "how does your team approach prototyping?" or even "why isn't your team using gears on the wheels?" While there is a lot of information on the Internet regarding FTC, it is scattered in many places, and frequently does not get into the fine details that can mean all the difference between success and failure.

The Pope John teams enjoyed a great season in 2009/10, with Team 248 winning the New Jersey Championship and Team 247 winning the New York City Championship, both as captain teams. In the 2010/11 season we added a third team, 4391, and they qualified for Worlds by being on the winning alliance at the New York Hudson Valley Championship. We've definitely learned a lot about what it takes to be a winning team at the highest levels of competition. What we've learned has come as a result of frequently painful experience, trial and error, and sometimes just plain luck. This book is an attempt to summarize everything we can think of to help new teams or even experienced teams to conquer the steep learning curve that this very complex and rewarding sport presents.

To gain even more perspective, we have sent enquiries to other championship teams, asking them to give us as many helpful hints as possible, and we have incorporated them into this text.

And before anyone asks, we have held back nothing! Everything we could think of, from the grand to the minute, is included in this book.

Past vs. Future Games

We should be clear, however, that everything in this book, while representing the opinions of several championship teams with proven track records in past seasons, are still just that–opinions based on past seasons. The FIRST organizers want to keep teams out of their comfort zones, they want to defy pat formulas for success, and they are very likely to throw numerous curve balls into the rules of future games.

So you need to consider carefully whether any particular tip applies to whatever game comes in the future, not merely whether it worked well in the past. These tips can help you avoid many common mistakes and give you food for thought, but they cannot substitute for your own analysis of precisely how the game rules differ from season to season.

In short, you should carefully consider the material we present here, but ultimately you need to use your own brainpower! Isn't that the entire purpose of the FIRST competitions?

Your Mileage May Vary

Can any team at all become a state champion by following the advice in this book? Well, if every team followed these tips, they wouldn't *all* be able to win their state championship; that would be impossible!

But what is possible is for every team to improve based on this book's advice, raising the level of competition generally, and leveling the playing field by helping close what we perceive as an "information gap" between the top flight teams and others. What we have seen is that the difference between an average team and great team often comes down to what may seem to be trivial points.

We have found, much to our horror, how the most minor mistake can lead to sudden defeat. Being aware of common mistakes, being vigilant in training to recognize and compensate for different kinds of mishaps that are bound to occur, and thinking ahead of all the possible failure modes will certainly help improve your team's performance. You are still subject to the vagaries of hardware and software; there is always some element of luck. Even the best teams sometimes lose because nobody can control every crazy thing that can happen out on that field. That's what makes it interesting! But you can put yourself in the best possible position to overcome difficulties. The best teams consistently do this; they make their own "luck" by doing everything possible to succeed.

And let's also not lose sight of one very important fact: corny as it may sound, winning really isn't everything. The FIRST competitions are primarily about inspiring students. Many of the items in this book will bring you much further along the learning curve with less frustration, and that has to be better for inspiring students.

That said, hey let's face it, it is fun to win at least *sometimes*! So good luck out there!

Tip, Trick, Strategy, or Secret?

This book is composed of a series of sections that group together related tips, tricks, strategies, and secrets that are commonly used by champion teams. So you might ask, what is the difference between a tip, a trick, a strategy, and a secret?

Well frankly it's somewhat arbitrary, but generally speaking here's how *we're* defining these terms:

- TIP: a specific, concrete, "helpful hint" that may be obvious to some team members and not obvious to others.

- TRICK: still specific and concrete, but less obvious than a tip, something that many or even most teams are probably not aware of, but offers a simple way to improve performance.

- STRATEGY: less specific than a tip or trick, a strategy is more like a "rule of thumb." It is more general and may be applied different ways in different situations. For example, a specific technique for mounting a gear would be a tip or a trick, while a more general item such as "design your robot to be heavy" is a strategy since there are many different ways to make a robot heavy.

- SECRET: this is reserved for those items that in our experience very few teams have figured out, but which are critical for success. They may be specific or general. Fewer than 10% of our tips fall in this elite category. Pay special attention to these items: they mean the difference between winning and losing at the very highest levels of competition.

Your team will, more than likely, already know some or perhaps even many of the tips in this book, but we're confident that you will still find many nuggets that help you on your journey through the FTC competitions. Even if your veteran team members are already aware of some or most of what's written here, you may find this book presents a great way to bring new team members up to speed on both basic and advanced concepts that are crucial for success.

Using This Book

The ways of using this material are probably as varied as the many strategies and mechanisms you'll see during competitions. It would be useful to read through at least the one-sentence description of each tip to get an overview of all the material. This can be done at a single sitting. Then later review in more detail applicable sections based on what stage you are at in the season.

For example, carefully read the section on brainstorming and prototyping as soon as the new game is announced to help guide your design process. Have the build team review general building techniques, and subteams in charge of the chassis, drive train, etc., review their own sections carefully. When you have a robot up and running for the first time, review the sections on driver training, drills, and practicing, autonomous programming, etc., then review the sections with hints on the Tournament before your first competition.

It can't hurt to have everyone read through everything—of course!—but practically speaking there is little reason for the hardware team to learn a lot about the programming side of things and vice versa.

You may find times when you face a certain problem or are trying to debug a hardware or software issue, and someone says, "Hey wasn't there a tip on this in the book?" Then you can look it up and we hope that saves you some time and effort in solving whatever problem confronts you.

Acknowledgments

This section lists all students and teams who contributed to this book, along with the nature of their contribution. If you'd like to see your team and your name listed here, all you have to do is contribute material to us! If we use your material in the next release of this book, you'll get attribution right in this section. So what are you waiting for? You can email your favorite tips to pj.ftc.tips@gmail.com. Be sure to also email supporting photographs, diagrams, or other materials to help explain and illustrate your tips.

Contributors to the Third Edition:

- **TEAMS 247, 248 Pope John XXIII Regional High School, Sparta, NJ**
  - Caitlyn Cherepakhov: Editor-in-Chief
  - Sonia Geba: Index, section introductions, new tips
  - Stephanie Warsh: Tips, section introductions
  - Jacquelyn Elio: Photographs
  - Jack Gilbert: Photographs
  - Jacob Scordato: Cover illustration

Contributors to the Second Edition:

- **TEAMS 247, 248, 4391 Pope John XXIII Regional High School, Sparta, NJ**
  - Emma Kelly: Editor-in-Chief, Scouting and Team Spirit Tips
  - Jacob Scordato: Cover illustration, Photographs
  - Michael Henning: Programming Tips, Code samples, Code repository, FCS Tips, Circuit diagrams
  - Max Bareiss: Programming Tips, Code samples, Code repository
  - Joseph Vengen: Main interior illustrations, Tips
  - Caitlyn Cherepakhov: Copy editor
  - Jacquelyn Elio: Additional illustrations, Photographs, Tips
  - Dylan Wynne: Photographs
  - Some additional material contributed by: Patrick Murphy, Joseph Link, and Hans deWaal
- **Team 4211 "The Bombers" John Burroughs School, St. Louis, MO**
  - John Stegeman: Ultrasonic sensor tips, motor encoder tips, miscellaneous programming tips, and code examples

Contributors to the First Edition:

- **TEAMS 247, 248 Pope John XXIII Regional High School, Sparta, NJ**
    - Emma Kelly: Editor-in-Chief
    - Ryan Bloodworth: Photographs, copy editing
    - Ryan Tarnopoll: Photographs
    - Jonathan Popo: Photographs, Tournament Tips
    - Michael Janov: Photographs, Building Tips
    - Victor Kaiser-Pendergrast: Code samples, Photographs, Bluetooth Appendix
    - Kati Hein: Marketing
    - Michael Henning: Programming Tips, FCS Tips
    - Jake Scordato: Cover illustration
    - Joseph Vengen: Interior illustrations
    - Max Bareiss: Programming Tips, Code samples
    - Hans deWaal: Building Tips
    - Zachary Zavoda: Building Tips
- **TEAM 3539 "Say Watt?" Edison, NJ**
    - Frankie Carr: Tips and suggestions
    - Pronoy Biswas: Tips and suggestions
    - Jon Silvestri: Tips and suggestions
    - Kaushal Parikh: Tips and suggestions
- **TEAM 3415 Lancer Robotics, Livingston, NJ**
    - Safety tips; some other miscellaneous tips.

Part I: Team, Meetings, and Engineering Notebooks

This section contains tips about organizing your team, including meetings, new recruits, communication, and even tips for organizing paperwork. In addition, since the Engineering Notebook is a crucial aspect of FTC robotics that must be attended to early in the season, we cover it here.

Safety is FIRST

Safety is the most important consideration for any robotics team. So what topic is better to begin our list of tips?

Before we begin, though, be sure to read all manufacturer safety information carefully for every tool, material, robot kit part, battery, etc. that you use. **If anything in this book conflicts with manufacturer instructions, always follow the manufacturer's advice instead.** This book contains general advice that may or may not apply to specific situations, so always use common sense and think through your actions before proceeding with any cutting, drilling, wiring, or other process.

If everyone observes sensible precautions there should never be any injury more serious than an occasional small scrape, pinch, or scratch, and even those should be rare.

 # 1. TIP: Every team member should wear safety glasses at all times.

Even if one student is not using any tools, someone around him or her might be, and during mishaps shards of metal or plastic may fly long distances. The glasses only come off when the meeting is over and all the tools are put away.

 # 2. TIP: When using loud power tools, ear protection is also required.

Follow the power tool manufacturer's advice on when ear protection is necessary.

 # 3. TIP: Adult supervision is required, especially when power tools and electrical systems come into play.

Students should never be alone in the workshop without supervision.

4. *TIP: Students must be trained on the proper use of tools, with an emphasis on safety, and pass a safety test before being allowed to use them.*

Using a bandsaw, drill press, grinder, etc. without proper training is just asking for trouble, but even common handtools can cause injuries if not used properly.

5. *TIP: Students who observe their peers acting unsafely should speak up, to mentors if necessary.*

Mentors can't be observing everything all the time, so students need to stay alert. Don't be afraid to speak out if your peers are acting unsafely.

6. *TIP: After metal or plastic parts are cut or drilled they must be filed to insure there are no sharp edges or burrs.*

This is not only a good idea to avoid getting sliced later, but you can also fail hardware inspection if you have sharp edges.

Emory paper can also be useful for dulling metal or plastic in hard to reach places.

7. *TIP: Wear appropriate clothing when working: no open toe shoes, no dangling jewelry or other items that could become caught in machinery, etc.*

Items sometimes get dropped, and your shoes offer crucial protection against injury from most of the hand tools and other materials that are used in FTC, but only if they enclose your entire foot. Loose jewelry may look nice, but not when it gets caught in a whirling drill press or even the robot itself.

In the illustration, the student on the left is violating every rule of safe workshop attire: improper footwear, dangling jewelry, hair, and other accessories, no safety glasses, etc. The student on the right has tied her hair back and corrected the other problems.

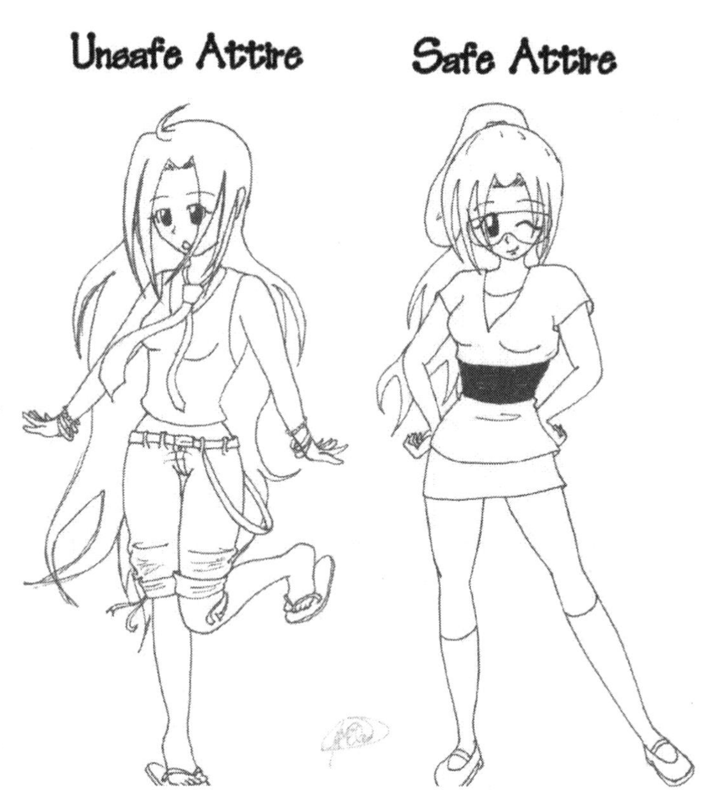

\# 8. *TIP: The workspace must be kept uncluttered. Cords should be kept out of walking paths. Students should always be aware of what is going on around them.*

When tools and materials are piled up on the workbench you cannot find what you're looking for and you also increase the chances that something will get knocked onto the floor or get in the way of a construction operation. Keep the workspace clean. Clean up as you go during build sessions and put everything back in its place at the end of sessions.

\# 9. *TIP: Keep a first aid kit in the workshop in case of any minor cuts.*

They cost under $20 at any pharmacy. Read up on standard first aid procedures in the booklet that is usually contained in these kits.

\# 10. *TIP: Any accidents, no matter how small, must be reported to your coach or adult mentors immediately.*

Let the adults be the judges of what is a large or small incident. Even a tiny cut, if not properly washed out and bandaged, can lead to an infection.

\# 11. *TIP: A telephone should be available in the workshop in case of a more serious injury.*

In the age of cell phones this is not usually a problem, but if cell reception is poor in your location it's a good idea to have a land-line in the workshop.

 # 12. TIP: Avoid "horseplay."

When a person is near or touching the robot do not actuate any effector or drive motor as a "joke." The motors are stronger than you think, and some mechanisms could seriously injure someone if a finger or other body part was in the wrong place. A thirty pound metal robot running into your leg at full speed or a motor driven metal dump bucket banging into your head is no laughing matter.

Apart from the robot, pushing someone, running in the workshop, throwing things, etc., is more dangerous than you might think. Don't fool around!

 # 13. TIP: Always turn off the "kill switch" or unplug the external battery before doing any repair or adjustment to the robot.

This reduces the chances of shorting out and possibly destroying electronics, and avoids motors being accidentally actuated when fingers are in the way.

Things Every New Team Member Should Know

 # 14. TIP: Make sure all team members learn and use the correct names for all the Tetrix, Lego, Hightechnic, and other parts used to build the robot.

This seems very simple but it is quite important and frequently not practiced. There are many ways this helps your team. You will save huge amounts of time in communicating with each other. Saying, "Please hand me an L-bracket," is far more efficient than saying, "Please hand me one of those little bent metal things. You know, the short ones. No, not that one, the other one! No, no, the *other* other one!" (That's an actual conversation from one of our meetings!)

You should especially use the proper terminology in your engineering notebook, and when speaking to the judges at tournaments. This makes for a far more professional presentation.

You can see all the part names over on the web site you use to order them: parts.ftcrobots.com for Tetrix parts, Lego.com for sensors, etc. Print out the catalog pages and distribute copies to all new team members each year and make sure people use those names. Your coach may even consider quizzing new members to ensure they know and are using the proper names for all parts.

This also goes for materials like screws, nuts, washers and even tools. We've seen lots of cases where a student asked for a screw but really meant a nut! Asking for a T-handle nut driver will get you the right tool quickly, while asking for "one of those long tightener thingies" will just waste time.

 # 15. TRICK: Intelligently name the structures on your robot, and consistently use those names in discussions, engineering notebooks, judging presentations, and computer code comments and variable names.

This is similar to the TIP above and is just as important though not quite as obvious. When you start to prototype and construct your robot each season, give intelligent sounding names to the major structures. Use names that make sense or are analogous to other similar machines. Use these names in casual discussions, brainstorming, the engineering notebook, judging presentations, etc.

For example, let's say you have some metal structures holding up a hopper that stores game elements. When you're talking to the judges at a tournament, instead of referring to them as "these hunks of metal over here," wouldn't it sound far more professional to say, "these hopper support brackets," instead?

As an example, in the 2009/10 season we constructed a mechanism to shoot wiffle balls into goals. Composed of a PVC pipe with a hole cut in one side to admit the balls, it contained a smaller diameter PVC pipe inside that was propelled toward the ball by a stretched surgical tube. The small tube inside had metal on the end to strike the ball. We named the outer PVC pipe the "barrel" in analogy to a gun. We named the inner tube the "piston," and the metal that struck the ball the "hammer." The hole cut in the barrel was called the "breach," again in analogy to a gun. There was also a servo motor that moved another piece of metal in front of the breach to stop balls in the hopper from pressing against the ball about to be fired. This was named the "breach guard." These terms made sense, were easy to remember, and made communication between both team members and judges easier. Using the same words when speaking will also give your

team a desirable appearance of cohesiveness and organization. The very same terms were also used in the notebook.

Some terms for major robot structures are so common you should always use them, and we will use them in this book:

- Chassis: (pronounced CHA-see) the base frame of the robot on which other items are mounted, usually rectangular and made from strong channels.
- Mechanism: any structure that has moving parts.
- Actuator: a part that causes motion under power; in FTC typically driven by a 12v motor or servo, but pneumatics can also provide actuation.
- Drive train: the combination of motors, wheels, gears, chains, etc. that make the robot move over the floor.
- Hopper: a container on the robot that stores game scoring elements like pucks, balls or whatever is being used in the current season's game.
- Harvester: a mechanism used to convey game elements from the field into the hopper.
- Indexer: a mechanism that helps feed game elements from a hopper into some kind of scoring mechanism such as a shooter. The act of feeding a game element into position to be scored is often called "indexing."
- Agitator: a mechanism that prevents game elements from becoming jammed in the hopper or while being indexed, generally by shaking or moving the game elements so they don't get stuck.
- Effector: a general term for some mechanism on the robot that performs some specific task, typically using servos or motors, for example an arm that bends to pick something up, a clamp that closes on a puck, etc. Effectors that are affixed at an extremity of a mechanism are called end-effectors.

 # 16. TIP: Teach each student the immortal phrase: lefty loosey, righty tighty.

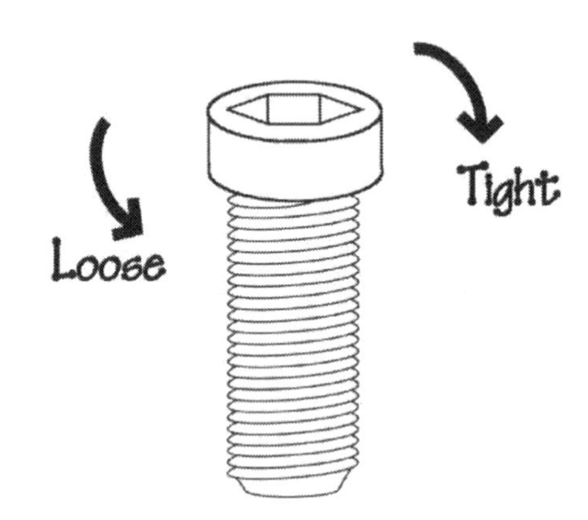

It's amazing how few kids these days know which way to turn a screw or nut to tighten it versus loosening it! Every season we see some student spend five minutes trying to remove a stubborn nut only to find they were turning it in the wrong direction the entire time. Drill this phrase, as well as a solid understanding of left and right, into new team members every season. A similar memory device is: Clockwise Closes.

 # 17. TRICK: Using colored tape, color code small hex keys that are not marked with their size.

 # 18. TIP: Drill students to know which hex key size (color) is used for which parts.

These two tips go together, and they are once again all about saving time and frustration when building. Time is precious, and spending several minutes of trial and error finding the correct sized hex key for a given part wastes that valuable time. A small strip of colored tape wrapped near the center of hex keys that are not size marked is all it takes. For larger T-handle or otherwise marked tools the tape is not necessary but still might be visually useful.

Here is a quick summary that students can study for just a few minutes at the start of season. Just recite the short mnemonic phrases shown in parentheses: White Collar, Blue Button, Green Hub, Red Socks (sockets), a few times. Then quiz yourself the next week, and you will know them by heart in no time.

SIZE	COLOR	USED FOR:	MNEMONIC	PICTURE
1/16"	White	Set screws used in axle set collars and motor encoder.	White Collar	
5/64"	Blue	All #6-32 button head cap screws.	Blue Button	
3/32"	Green	Set screws used in axle hubs and motor shaft hubs.	Green Hub	
7/64"	Red	All #6-32 socket head cap screws.	Red Socks (sockets)	

Meetings

Practice and communication are essential to success. These are what team meetings produce and facilitate. It is important to set aside time to make sure everyone on the team is on the same page concerning everything from robot design to scheduling of tournaments. Bottom line: meetings are vital to a team's success.

 # 19. TIP: Maintain an accurate team email list, including parents, and clearly communicate meeting schedules.

Team communication is essential and email is a great way to facilitate it. Meeting notices should go out weekly, at least a day or two before the meeting. A short description of the purpose of the meeting is helpful, it need not be elaborate. Keep emails brief and to the point.

 # 20. TIP: Keep most meetings in the range of 2 to 3 hours.

Meetings that are too short don't allow students to get organized and accomplish a solid amount of work. Meetings that are too long sometimes result in many students losing focus. The exception to this tip is that when you get very close to a tournament, especially an early scrimmage when the robot barely functions, you may need to have one or two "marathon" sessions to get everything nailed down in time. Some students may have downtime during such long sessions as they are stuck waiting for other students to complete some task on the robot (for example the programmers may have to wait for a hardware fix). It's a good idea to urge students to bring homework or reading assignments so they can make such downtime productive.

 # 21. TIP: Meet at least 5 hours per week starting immediately after the new game is announced.

Over and over again we will learn in this book that time management is a key factor in a Champion team's success. We have been at many tournaments where we hear a coach lament, "We *almost* got our scoring system perfected; we just needed a few more build sessions!" While many of the tips in this book have to do with saving time and making meetings more productive, there is no getting around the fact that a typical champion level robot requires a minimum of 100 hours of work by 5 to 10 students. That's 500 to 1,000 engineering hours to build a championship level robot.

So do the math: if the game is announced in the middle of September, and your championship tournament is in the middle of January, then you'll need 100 hours of meetings in 18 weeks, or about five hours per week with a few marathon sessions thrown in.

If you are in a state that has a much earlier championship, well, every other team in your state is in the same boat so we're not suggesting you meet 10 hours per week to compensate for the shorter build time, but you may need to work a more rigorous schedule than suggested here in that case.

Typically we have a 2 hour meeting on a weeknight (being sensitive to student's need to do homework) and a 3 hour meeting on the weekend, with one or two marathon weekend sessions near the first scrimmage and again near the first major tournament. We usually omit the weekday meeting following a tournament to let everyone recuperate, and we take off Christmas week and Thanksgiving week. You will, of course, adjust this based on your team's needs.

 # 22. TIP: Not every meeting needs to involve every student.

Especially late in the season, it is often more productive to split up meetings, such as have one meeting that is only for autonomous programming, or one meeting that is purely for driver training. While it is good to have a couple of hardware specialists at those meetings in case a fix is needed, there is no reason to have everyone show up for a meeting that is primarily going to involve the programmers or drivers only.

After the robot is basically sound and running, we will sometimes make the weeknight meeting a specialized session for a certain purpose that only involves a subset of our team (e.g. autonomous programmers or drivers), but usually the weekend meeting will continue to be a general build session that may involve several different teams.

 # 23. TIP: Most students should specialize in a handful of roles.

At the 2009/10 kickoff meeting, we heard one new coach state that he wanted every student to learn every aspect of the robot, including programming, driving, hardware, electrical systems, etc.

We found that this is not the best idea. First of all, not every student wants to be a programmer, or a driver. Second, there simply is not time in a build season for everyone to learn everything. As new coaches quickly find out, there is a gigantic amount of information necessary to build a successful FTC robot. Students will need to specialize, and fairly early.

Certainly there will be some students who excel at several different major tasks, but that is not typical. It is important to utilize each student's individual skills in order to most efficiently benefit the whole.

This does not mean students should be barred from knowledge outside their specialties—a smattering of knowledge across the spectrum of skills is desirable. But students will probably only have time to go into depth in a handful of areas.

Regarding driving, there typically is not enough time to allow every student to intensely train to drive the robot. Practically speaking, the robot is usually barely functioning by the time of the first scrimmage, and it takes many hours for drivers to gain enough experience to outmaneuver opponents on sometimes shaky hardware, learn through trial and error all the fine points of the new season game, etc. About the best you can hope for is to intensely train one primary set of drivers and one secondary set. We have try-outs where performance on specific tasks (driving, harvesting, scoring) is evaluated to pick the drivers in the first scrimmage.

Organization

 # 24. TIP: Encourage parents to get involved.

FTC Robotics is the kind of sport that benefits greatly from parental support because there are so many different aspects to the program. Parents who have technical careers such as engineering or software might become great mentors for the team, but there are many non-technical aspects such as public speaking to judges and sponsors, pit decorations, marketing, publicity, and fundraising that can be improved by adult guidance. Enthusiastic and knowledgeable as they may be, just having several school faculty involved to advise the team may not be enough for a well rounded program. Parental support can be crucial for success.

 # 25. TIP: Create a team handbook that outlines all team responsibilities, rules, and procedures.

A team handbook ensures everyone, including parents, knows basic information about how the team operates and what the rules are. See the Sample Team Handbook in Appendix E, although your handbook may be quite different depending on your needs and policies.

 # 26. TIP: Inventory enough parts; avoid expedited shipping charges by keeping on top of inventory needs.

Often during competitions or testing you will damage various parts of the robot: motors burn out, sensors get damaged, etc. By ordering in larger quantities you can insure that you always have a specific part on hand for immediate repairs, or to try out new ideas when building. Besides being able to repair your robot on the spot, you won't have to pay for express shipping to get in repairs before your next tournament. Better to spend money on hardware, not shipping!

 # 27. TIP: As much as budget allows, order at least one of every available part.

Over the season you will make changes to your robot and it is a good idea to have different materials to experiment with, without waiting for them to be shipped. This is important because very often there are multiple approaches to a problem and it is preferable to be able to test out several to find the approach that works best. By having all parts on hand you can try out your ideas immediately.

The Engineering Notebook

 # 28. TIP: Read and understand the guidelines for good notebook pages presented in the Game Manual before making any entries.

There are very specific rules outlined there, such as that all notebook entries should be in pen and not pencil, that each page should be signed and dated, that there should be no blank pages or large areas left blank on any page, and other crucial information. Many of these rules are the same rules that real corporations use for engineering notebooks, so you are learning real procedures by following these rules.

 # 29. TIP: Train new students on correct notebook entry techniques; have them do some practice entries before writing in the real notebook.

Students who are new to robotics may need some instruction and practice before making their first real notebook entry in the official notebook. Remember, once something goes in the notebook it never comes back out, so it's better to have new students practice by making sample notebook pages on sheets of paper. Then, have veteran team members or mentors tell them how they could improve those first efforts before transferring the information to the engineering notebook. After a few practice pages, students can start writing directly in the real notebook.

 # 30. TIP: Everyone on the team should make entries in the notebook.

Some teams do designate one team member as the notebook coordinator, but that doesn't mean that one person has to write every entry. We prefer to make sure that every team member makes at least some notebook entries. We have had some judges comment that they liked that about our notebook, they preferred to see the entire team participate in the notebook. Now, this does not necessarily mean that every team member will contribute equally to the notebook. Some people are just naturally more talented at writing and illustrating. But everyone should make at least some contributions.

 # 31. TIP: Keep a digital camera, printer, and glue stick in the workshop.

This allows your team to take pictures, print them, and glue them in to illustrate notebook entries immediately. A low cost paper trimmer can also add to the professionalism of such photographs, you can trim using this more neatly than scissors allow. Remember to draw a box around any pictures glued into the notebook using pen.

 # 32. TRICK: Set up an area in the workshop specifically for taking photos of mechanisms, it should have good lighting and a white cloth or felt background.

Along with the previous TIP, some pictures may be harder to take than others. While you may be able to snap a photo of a large mechanism without any special setup, small parts may come out better if you have a specific place set aside for photography in your workshop. It need not be elaborate. A nice white background, either felt purchased from a fabric store or just a plain white cloth or even a piece of poster board will do. A desk light to ensure lighting is sufficient is also helpful.

 # 33. TIP: A great place to get engineering notebooks is www.bookfactory.com (BOOKFACTORY).

Many teams use this source for professional notebooks. In fact, these are the very notebooks used by real engineering companies to record their results and back up patent applications!

 # 34. TIP: Purchase the large size notebook (8-7/8 by 11-3/8 or larger) because it's easier to glue in program code print outs and other materials printed on standard paper.

Several different sizes of notebook are available. The larger ones will reduce the amount of trimming you have to do to insert photos or print outs of your autonomous or tele-op computer code.

 # 35. TIP: Have your team name and number engraved on the cover.

It doesn't cost much and adds to the professional appearance of your notebook. It also helps you find your notebook when half the teams at a tournament use the same kind!

 # 36. TIP: Make entries at every build meeting. Mentors might consider prompting students to start making entries at least half an hour before the end of the meeting.

Don't put off until tomorrow what you can do today! And that goes for engineering notebooks. While the ideas are fresh in your mind, make the entries at the same meeting you did the work.

 # 37. TIP: Reserve 10 to 15 pages at the end of the notebook for outreach activities.

Some teams put their community service and outreach activities at the end of the notebook, separate from the build meeting notes. We have found this to be useful.

Make sure you mark such sections so judges find them easily, using post-it notes or similar markers.

 # 38. TRICK: Document failures as well as successes.

If you are really stretching your imagination and trying many things, you may actually have far more mechanisms you build fail than succeed. This is the real world of engineering! Document those failures just as precisely as your successes. We have had judges comment that they were impressed how we documented our (many!) failures.

 # 39. TIP: Consider making an electronic notebook instead of a traditional paper notebook.

For the past two seasons our team has been making our notebooks electronically instead of the traditional paper notebooks. There are very specific rules about how to do this, so read the game manual carefully and make sure you are following the rules. Some big advantages of an electronic notebook are: easy incorporation of pictures and diagrams, easy incorporation of color, more legible than handwriting for the judges to read, and more.

Be careful though: when using an electronic notebook there may be temptation to go back and edit past pages. The procedure is to write up the day's information, then print out the pages and sign and date them. They are then inserted into a binder and should not be modified after that.

 # 40. TRICK: Be visual. Try to include a picture, diagram, sketch, flowchart, etc., on every page.

Let's face it; the judges only have a very limited amount of time to evaluate your notebook. So it's important to grab their attention. Nothing does that as well as graphics. Pictures, sketches, diagrams, flow charts, tables of test results, these are all things that will vastly improve your notebook pages. Strive to have almost every notebook page contain some striking visual element.

The first notebook page shown below uses pictures effectively to illustrate a complex revamp of our 2009/10 harvester mechanism. The second notebook page shown uses a combination of photographs and circuit diagrams to illustrate how a new Prototyping Board was constructed.

Trying to describe the same information without pictures would have been both difficult and far less interesting and informative for the judges to read.

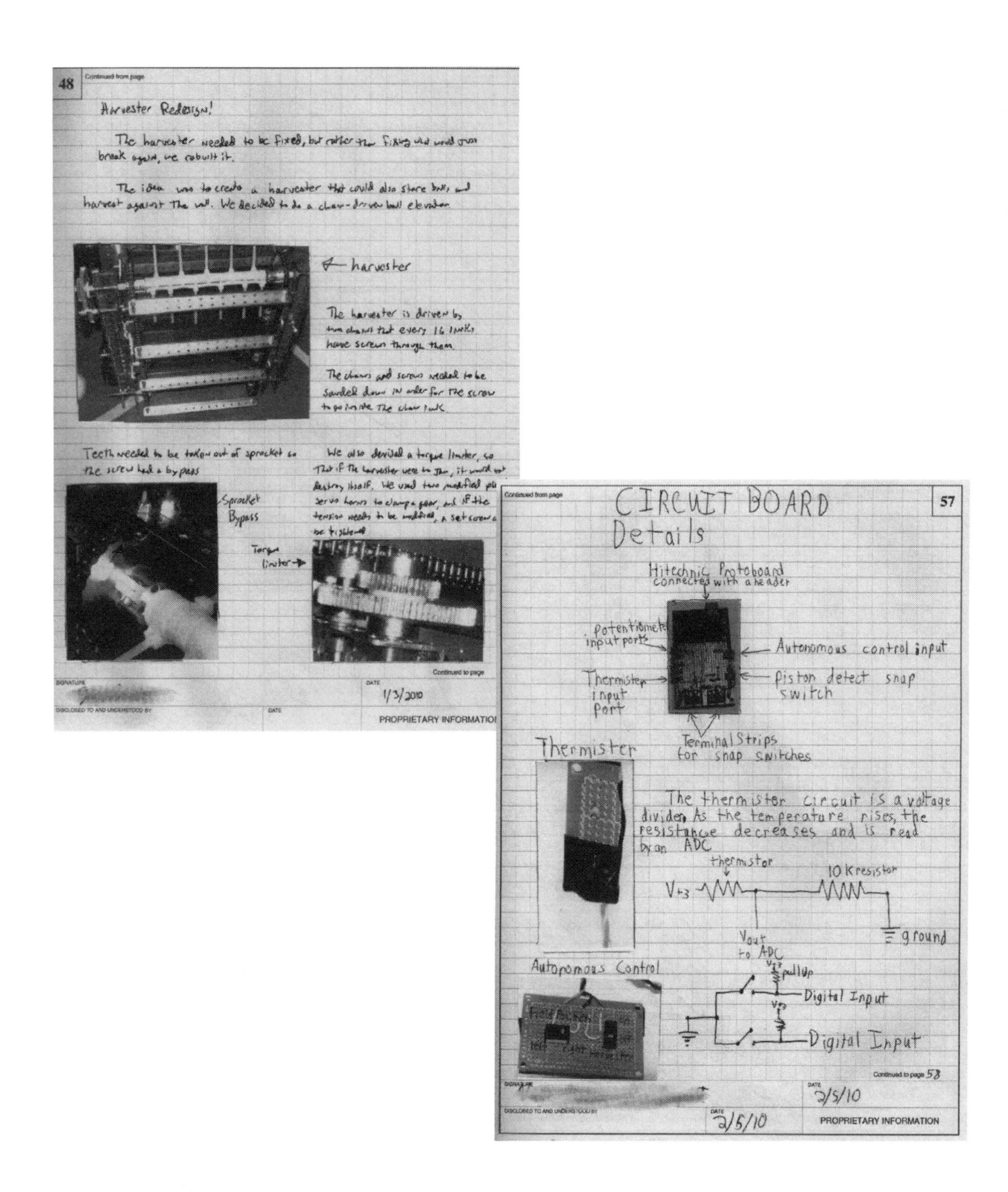
48
Continued from page
Harvester Redesign!
The harvester needed to be fixed, but rather than fixing what would just break again, we rebuilt it.
The idea was to create a harvester that could also store balls, and harvest against the wall. We decided to do a chain-driven ball elevator.
harvester
The harvester is driven by two chains that every 16 links have screws through them.
The chains and screws needed to be sanded down in order for the screw to go inside the chain link.
Teeth needed to be taken out of sprocket so the screw had a bypass
We also devised a torque limiter, so that if the harvester were to jam, it would not destroy itself.
Sprocket Bypass
Torque limiter
Continued to page
SIGNATURE
DATE
1/3/2010
DISCLOSED TO AND UNDERSTOOD BY
DATE
PROPRIETARY INFORMATIO
Continued from page
57
CIRCUIT BOARD Details
Hitechnic Protoboard connected with a header
potentiomete input ports
Autonomous control input
Thermister input port
piston detect snap switch
Terminal Strips for snap switches
Thermister
The thermister circuit is a voltage divider. As the temperature rises, the resistance decreases and is read by an ADC
thermistor
10 K resistor
V+3
Vout to ADC
ground
pullup
Digital Input
Digital Input
Autonomous Control
Continued to page 58
SIGNATURE
DATE
2/5/10
DISCLOSED TO AND UNDERSTOOD BY
DATE
2/5/10
PROPRIETARY INFORMATION

Part II: Brainstorming, Design, and Prototyping

When the new game is announced each season, typically around the middle of September, there is excitement in the air. Everyone on the team will likely have their own ideas about the game and most students will want to jump right into building. But it's best to have at least a couple of meetings where students "brainstorm" different ideas, followed by sketches, prototypes, and some testing to weed out all but the most workable solutions. This section has tips on how to manage the brainstorming, prototyping, and initial design process to maximize your chances of coming up with a world class robot.

The Brainstorming Process

 # 41. STRATEGY: First discuss the game rules extensively without referencing robot design at all.

If everyone does not have a very good understanding of the game you are bound to go in the wrong direction. Typically FIRST releases a short video outlining the basics of the game, but the real meat is in the Game Manual. There are often very important details in the Game Manual that are only briefly mentioned, and sometimes not mentioned at all, in the video. We typically review the video several times, and then mentors have the students read the most important chapters in the game manual between meetings.

 # 42. STRATEGY: Second, discuss general strategies for achieving different game objectives without referencing specific mechanisms.

Once everyone understands the rules, discuss strategies without deliberating on how a robot would actually achieve that strategy. For example, in the 2009/10 season you could score 5 points getting a ball into a high goal. Two strategies for doing this are: shooting the ball through the air or elevating the ball up to the goal without shooting it. These are strategies that do not reference specific hardware. There are many ways to shoot a ball through the air (catapult, slingshot, crossbow, etc.) and there are many ways to elevate a ball (scissor lift, arm, conveyor belt, etc.) Forget about the "how" part for now. Every time a student starts talking about a specific mechanism remind him or her that "we're not talking about mechanisms now, we're talking about general strategies." (Although such ideas should be jotted down somewhere so they are not lost later.)

The reason for doing it this way is that if you start talking about mechanisms too soon, you may miss the best solution. You may become intellectually "invested" in a specific proposal before you've thought out how all the different parts must work together.

Do this for all major aspects of the game: harvesting, scoring, defense, drive train, special parts of the game that are wildly different from year to year. Important: try to make a comprehensive list of every possible way you can imagine to solve each game problem, but without going into specific mechanisms. Even seemingly outlandish ideas may end up being useful, for they may suggest more practical lines of thought later. At this stage, nothing is rejected unless it is clearly impossible ("time travel back before autonomous mode and then…") or clearly against the game rules ("lay a minefield next to the opponent's goal!")

 # 43. STRATEGY: Finally, list specific mechanisms that can implement each strategy.

Now that you have listed every possible strategy for each task in the game, it is finally time to think about mechanisms that can achieve each strategy. Again, try to be comprehensive at this point and do not reject any idea other than those that are clearly illegal.

If you had determined that there were five different general categories of ways to score, now for each and every one of those five try to think of every possible class of mechanism that could implement each strategy.

Evaluating Alternative Designs

At this point you have created a comprehensive list of every strategy and mechanism you can possibly think of to solve the problems the new game presents. This section offers tips on how to narrow this large list down to just those candidate solutions worthy of prototyping.

 # 44. STRATEGY: Evaluate the advantages and disadvantages of each possible mechanism.

Evaluate each and every mechanism you listed for attributes like:

- Speed: will this mechanism generally be faster or slower than others?
- Accuracy: how consistently does the mechanism achieve the desired result?

- Complexity: how hard will this be to build? Will it be hard to keep it working and properly adjusted?
- Size: will it be hard or easy to make this fit in the robot's required dimensions?
- Programming requirements: does this mechanism require sensors and programming that might be difficult to integrate, or will it be straightforward?
- Overall: considering all of the items you discuss, rank each mechanism for how likely it is to be the best solution.

Now clearly at this stage of the season you will only be able to take your best guess on these evaluations. The goal here is to pick several designs that seem the most promising, then put them to the test and prove which one is actually the best.

 # 45. TRICK: Aim high: give more consideration to mechanisms that can score the big points.

Most FTC games have several ways to score, some ways yielding more points than others. We frequently see robots that are designed only to score in "low goals" that yield few points. Champion teams always try to design from the start to score in all the goals, and are optimized to score in the highest point goals.

We've heard teams make statements like, "well we can only score in the low goal but we're really good at it so that makes up the difference in points." This is almost never true. For example in the 2009/10 season you got 1 point for a low goal score and 5 points for a high center goal score. You're just not going to beat a team that can pump points into the high goal no matter how efficient you are at scoring low goals. They have a five to one advantage! In that season, it would have been physically impossible for the low goal to even hold enough balls to outscore just five or six high goal balls; which any champion level team could achieve.

 # 46. TRICK: No limits: start out by trying to solve <u>all</u> the problems.

Related to the TRICK above, most games have numerous "problems" to solve. For example in the 2008/09 season you could score by climbing up a ramp at the end of the game. In the 2009/10 season you could score a special "doubler ball" at the end of the game. In the 2010/11 season you could try to balance your robot and rolling goals on a bridge for extra points in the final 30 seconds. In all these seasons you could gain some points by dispensing game elements during autonomous.

We frequently hear during brainstorming, even among our own teams, students saying things like, "well we could solve problem A or problem B, which should we go for?" That's the wrong way of thinking. The correct statement from that student should have been, "how can we solve *both* problem A and problem B."

Champion teams try to solve *all* the problems. They do not leave any scoring opportunity unexplored.

Now, even the best teams may not succeed in perfectly solving every problem. But they try, and they usually come pretty close.

 # 47. TRICK: Speed is good. Time matters ... a whole lot.

When evaluating different designs, always consider speed. We have seen teams with amazing mechanisms for scoring points which stun the mind with their elegance and mechanical dexterity. But frequently these great machines are just too slow; they don't end up making it to elimination rounds.

This goes for all aspects of the game: harvesting, scoring, dispensing, etc. Of course there is usually a trade-off between speed and accuracy (or between speed and power). You need to balance those elements and try to find the right design that maximizes your scoring potential. But if you start with a design that is guaranteed to be slow, there's nowhere to go. If you have a scoring system that is 100% accurate but can only launch one game element every 10 seconds, it's going to lose every time to a 50% accurate opponent who can blast out ten game pieces in the same time.

 # 48. STRATEGY: Simplicity is good.

Given two equally well performing designs, the simpler one is always better. If something goes wrong(and things always go wrong), fewer moving parts, failure points, and necessary repairs can make solving any problems much easier. The simple design also means easier access to parts, which can be very helpful when a quick repair is needed between matches. Simple designs are also less expensive to build since they usually require fewer parts. Many times students will proceed along very complex lines when a simpler solution performs just as well and can be built in half the time with half the parts. Always strive for simplicity.

 # 49. TRICK: Start with a basic, solid design, then incrementally improve it using sensors or refinements.

Frequently we start with a simple, basic design that does the job, then do baseline performance measurements, then refine the basic design using sensors or additional mechanisms that address specific failure modes. At each stage we try if at all possible to leave open the option of going back to the simpler design if the "improvements" turn out not to really help, as proved by testing and comparing with the baseline.

Sometimes an improvement unintentionally adds an additional failure mode or slows down the scoring rate—testing is the only way to prove it.

 # 50. STRATEGY: Be a multi-dimensional threat.

Too often you see robots that have exactly one way to score. For example, in the 2009/10 season you could score in a central high goal, a central low goal, or an off-field goal which could only be used in the last 30 seconds. Some teams optimized their robots for the off field goal only since that was the highest point total. The problem was that they were subject to defensive actions. If an opponent successfully blocked them for those last 30 seconds, they lost. Teams that could score several different ways presented multi-dimensional threats to their opponents, no one single strategy could defeat them, and they were harder to beat as a result. The winning captain team at the 2009/10 World Championship, Smoke and Mirrors, presented such a multi-dimensional threat and was able to consistently beat teams that optimized only for end game scoring.

 # 51. STRATEGY: Whenever practical, avoid single points of failure.

Always consider what the result would be if one single item on your robot failed. Would you be totally out of action? Or would it just be a minor irritation? Any single item whose failure means your robot might no longer be able to play the game is a subject of great concern. That item must be checked thoroughly before every round to ensure it is still functioning perfectly.

For example, we have seen times when a student used a single screw to hold a part on the robot. If that lone screw comes undone, the part will fall off! Putting two screws on the part avoids a single point of failure.

You cannot always avoid single points of failure due to practical considerations. There may not be room for two screws on a small part, and you only get a limited number of some parts, such as motors, according to game rules (restrictions vary year to year). But you should favor designs that don't have a single point of failure, all other things being equal. Because of the given limits on parts, you may need to make a trade-off decision.

As an example of this trade-off, one of our teams had a shooter that actually ganged together two of the 12v motors to draw back a piston. Either one of the motors could totally fail but the shooter would still function. Because the shooter was a highly essential part of the game (without it you can't score!) this was a robust design that avoided a single point of failure. Our other team decided to live with the single point of failure on the shooter so they could use an extra drive motor on the drive train. This gave them greater pushing power, useful for defense and offense. This is a design choice; there is no right or wrong answer. But, knowing they had this single point of failure, the team learned that they must constantly check that shooter motor before every round (checking for a burning smell for example). And the motor did burn up several times during practices and even right before a real round at a tournament. But by checking it constantly, the team ensured that it would not fail during an actual game.

Another strategy for dealing with an unavoidable single point of failure is to make sure the pit crew can quickly repair it. Team "Say Watt?" had a claw mechanism to pick up scoring elements in the 2009/10 season. This claw sometimes broke, but by training their pit crew using a stop watch to stress speed, they learned to replace it in under 90 seconds.

Either way, understanding single points of failure will allow you to make these trade-offs intelligently.

Prototyping

This is another lesson in preparation: prototype often. Prototyping is the act of creating a quick mechanism to prove out a concept. It need not be pretty: duct tape and cardboard may be the materials of choice! Be aware that in the real world most prototypes are "failures" in the traditional sense of the word. But from each failed prototype you learn something, so even these "failures" are merely stepping stones on the path to success. So no need to be discouraged; prototypes are an important part of your engineering adventure! Prototyping is fun and constructive and develops an affinity for creative tinkering, which can lead to genius discoveries in the design of a truly outstanding robot.

 # 52. TIP: Students, especially new team members, should spend some time just playing around with the parts.

This can take place even before the new season's game is announced, and it is vital for students to develop an innate understanding of what the parts are capable of doing or not doing, how they fit together, etc. After the game is announced, experimenting with parts with the game goals in mind can still be useful for visualizing solutions.

 # 53. During initial prototyping, don't worry too much about size or material limitations.

Instead, focus on getting something, anything, that actually works to solve a game problem. Once you prove a concept, you can then work on reducing the size or coming back into materials compliance and usually can figure out a way to do that later.

 # 54. TRICK: Try multiple prototypes for each subsystem, worry about hooking it all together later.

Generally speaking, you can figure out some way to bolt everything to the chassis. So, during prototyping don't worry about how it's all going to fit together, just try different subsystems by themselves. Have small teams (1 to 3 students) pursue several different ideas for each subsystem. For example, try two or three different harvester designs, two or three different scoring mechanisms, etc. Don't just pick a single concept after brainstorming and stick with that one doggedly.

 # 55. SECRET: Use two-dimensional cardboard models to quickly determine feasibility and dimensions.

Sometimes it is useful to create a two-dimensional prototype to work out dimensions and prove feasibility of an idea. All you need for this is a ruler, scissors or razor blades, some corrugated cardboard, tape, stapler, glue and maybe some screws or finishing nails to represent pivots.

The picture below shows an early two dimensional prototype for our 2009/10 shooter concept. The wiffle ball is represented by a disk cut to be exactly the same diameter. Everything is cut exactly to scale. Note how the hopper and hollow interior of a PVC pipe is created by building up a layer of cardboard and leaving empty negative space. The piston used to fire the ball is represented by a rectangle of cardboard inside this negative space. A few slivers of cardboard are used to hold things together. We even put a rubber band on the prototype to propel the piston, and simulated the motor that would pull the piston back with a shaft of cardboard pivoting around a wood screw.

This prototype actually worked quite well. Rotating the shaft by hand (you can see someone doing this in the picture) actually pulled back the piston, allowing the ball to fall into place from the hopper, then as the shaft continued traveling around the piston was released, and the 2-dimensional "ball" actually fired out the barrel by the force of the rubber band!

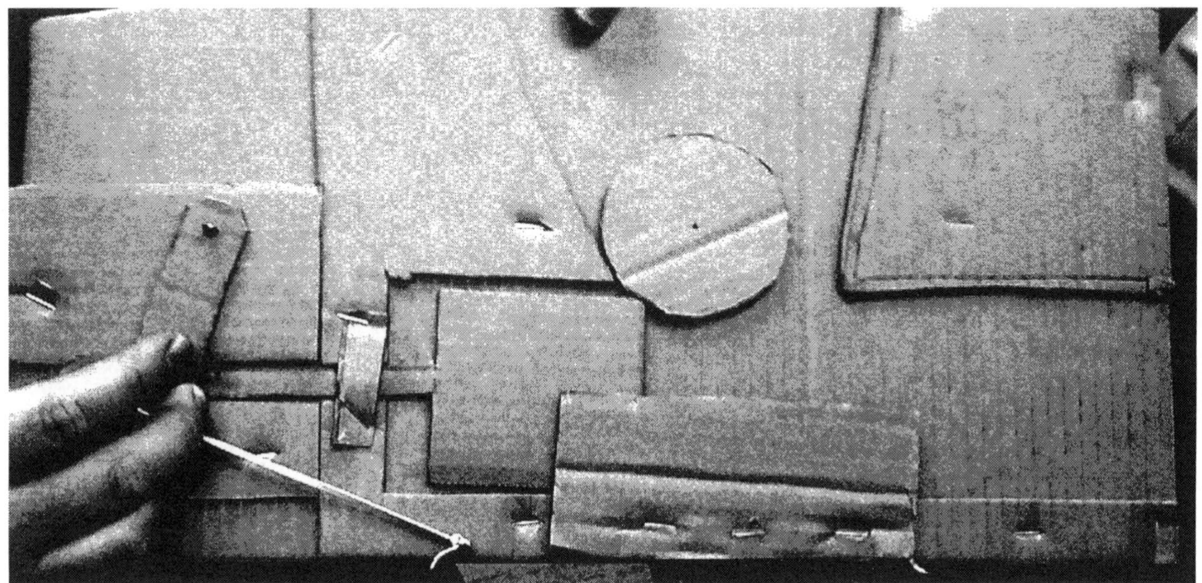

Because this two dimensional prototype was built precisely to scale, once it was completed and proven to work the exact locations of various holes and lengths of materials could be measured and easily transferred to the real PVC pipe and other elements. The motor position was known, the length of the shaft, everything. If we had just started hacking PVC to figure this out by trial and error it would have taken far longer, as we would likely have cut the holes in the wrong places for everything to work properly and would have had to re-cut the PVC several times to get it right.

A photo of the 2-D cardboard prototype is a great thing to include in the engineering notebook too!

 # 56. TIP: During initial prototypes, use tape, cardboard, and other easy to change materials before committing to metal and plastic.

Corrugated cardboard is an excellent analog for plastic but is far less expensive (free if you save old boxes) and can be cut easily with scissors or razor blades. Tape can be used to affix parts being tested, and using tape requires a fraction of the time it takes to fix screws in place.

Of course these materials are illegal for actual play and must be replaced before your tournament. But early in the evaluation and prototyping process you are simply trying to prove feasibility. You need to cut some corners so you will have time to try out everything, to get full evaluations of all your candidate mechanisms.

 # 57. STRATEGY: Objectively evaluate the prototypes before deciding which mechanisms to use.

Very often students (and adults!) get emotionally invested in a certain technical direction. To get the best mechanism, you need to set aside those feelings and use objective evidence as the only criteria for judging one design versus another.

For example, in the 2009/10 season, we tried many different shooter ideas before settling on one design that was clearly performing better in our tests. One student spent 15 to 20 hours investigating an alternative design. Although his design did work quite well and was an amazing mechanism, it was more complex and somewhat slower than the alternative design eventually chosen. This student looked at the evidence objectively, and regardless of the amount of time he had invested in his design he withdrew it from consideration voluntarily, recognizing that the alternative was superior. Regardless of the fact that the design was no longer going to be used, he meticulously documented in the engineering notebook the reasons why it was ultimately rejected. He demonstrated both a foresight and a selfless sense of the greater good that judges (and mentors) love to see. This is a student who was acting like a first rate engineer!

Using the Forums

 # 58. TIP: Monitor the forums on the FIRST web site.

The forums offer clarifications on building materials, game rules, allowed strategies, and more. To find them, just go to www.usfirst.org, click on the FTC link, then click on the FORUMS link (as of this writing on the left side of the page).

 # 59. TIP: In a notebook, keep track of the last date that you viewed the forums. Use that date to scan for forums that have new postings since you last looked.

The forums do not currently give you a simple way of seeing new postings since the last time you read them. Since new material is posted all the time, you need a way to quickly scan to see which categories (for example, Game Rules, or Build Materials) actually have new posts since the last time you viewed them. Fortunately, the forums do show the date of the last posting at the top level of the topic listings. Simply jot down the date that you last viewed the forums in a notebook. The next time you visit the forums, you need only read those categories that have had new postings since the last date you viewed them.

 # 60. TIP: Have your mentor post your own questions when necessary; but make sure you are not repeating a question already answered.

Your mentor will receive a login code with your team's registration that allows him or her to post questions to the forums. If you are unsure whether or not a particular strategy you are considering is legal, or have a question about whether a certain material can be used, ask your mentor to post the question. However, be careful to read and be familiar with all the related postings. Don't ask questions that have already been answered. Also, limit your posts to exactly one question per posting. Break up multiple questions into multiple postings.

 # 61. TRICK: If you intend on using an unusual strategy, and that strategy has been approved as legal in the forums, print out that forum posting and bring it with you to the tournament.

We have seen cases where referees or hardware inspectors were not 100% familiar with each and every posting to the forums. It is understandable: they are volunteers and there is a huge amount of detailed information posted on the forums with new posts constantly appearing, making it difficult even for referees to know every fine point. So, if you are planning to use a strategy which may be uncommon, and a forum posting has explicitly said that strategy is allowed, you will want to print that out as proof and bring it with you to the tournament. During hardware inspection, or at other times when the strategy or mechanism might be questioned, you can produce the print out of the forum posting that proves it is an allowable strategy.

Part III: Building

Most students join a robotics team, not surprisingly, because they want to build robots! Probably half of you have skipped right to this section just for that reason, even though you should remember that the preceding sections on team structure, brainstorming, prototyping, etc., are in many ways fundamental to building a solid foundation for success.

So fine, read this section first if you must! There's lots of good stuff here, to be sure. But don't forget that any champion values team structure and the beginning stages as an important part of the winning experience.

General Building Tips

 # 62. TIP: Don't cut metal parts unless absolutely necessary.

They are expensive so try not to cut them until you are sure you need to. During early building and prototyping it's fine to allow some pieces to stick out (as long as that doesn't stop your mechanism from functioning); you can cut them later to neaten things up.

 # 63. TIP: Measure twice, cut once.

This old proverb is a good one to take to heart. Make sure your marks are correct before cutting—parts are expensive and time is precious! It's also a good idea to have someone else take a look at your marks before a complicated cut begins.

 # 64. TIP: Use cardboard to prototype initial sizes of plastic and metal cut parts. Once precise dimensions are determined, use the cardboard parts to trace for cutting the real material.

As with your initial prototypes, using cardboard for the first attempts at the actual competition robot parts saves you time and effort. Often you will find out that you need to trim something here or there, and it's just a lot easier and cheaper with cardboard. When everything works the way you want, carefully remove those cardboard parts from the robot, trace them onto the final material (usually plastic or aluminum sheet metal), and voila, you have gotten the exact part you need in the most cost friendly and efficient way possible.

 # 65. TRICK: Keep the cardboard versions used as templates for plastic or metal cut parts for your "cut sheet."

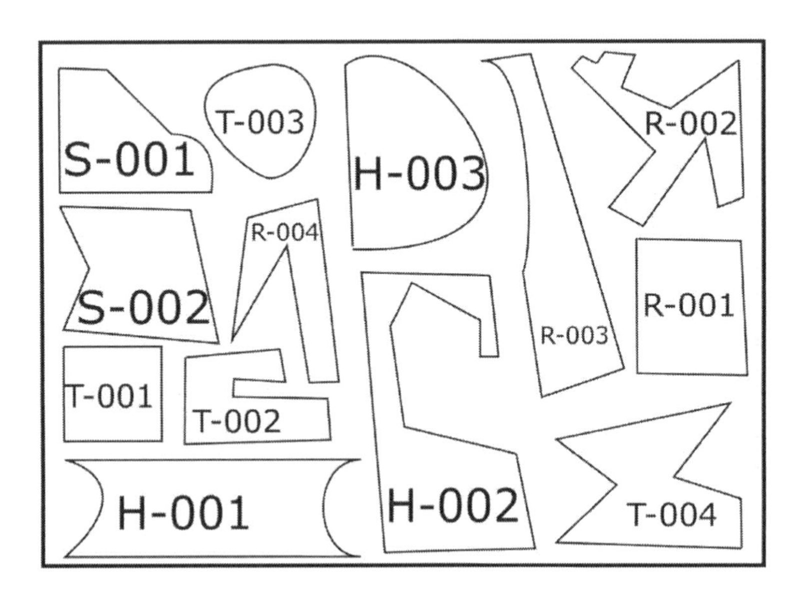

After tracing the part, don't throw the cardboard version away! Instead, keep a file of all those templates. You can use them later to create your "cut sheets," which will be used to get you through hardware inspection. A cost effective way to keep them organized is to purchase a cardboard "artist's portfolio" from an office supply store, for example the "Smead Red-Rope Artist Portfolio" which comes in sizes easily large enough to accommodate the part sizes common in FTC.

If you later create a new version of the part, mark the old cardboard template version OBSOLETE with indelible marker so you don't get confused about which parts are actually on the robot.

The illustration shows a cut-sheet as you might present it to a hardware inspector. It shows how you would cut each part out of a sheet of material of the size allowed by the game rules. There would be one sheet for each different material allowed.

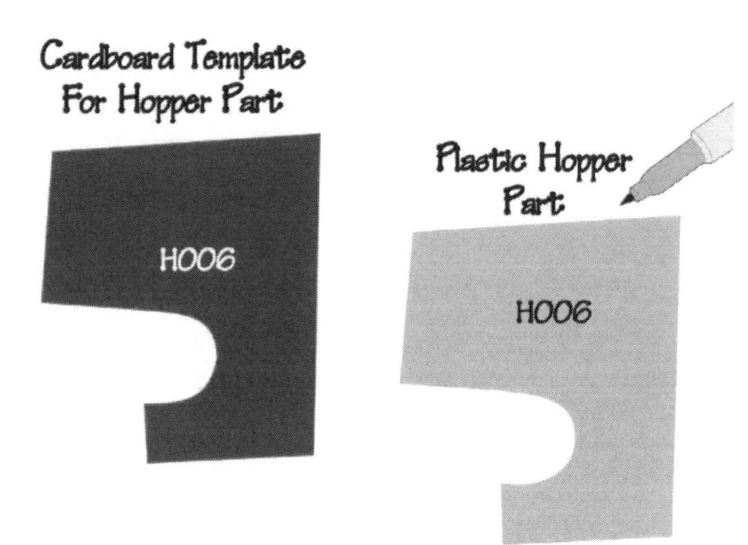

 # 66. TRICK: Use indelible marker to label each plastic or metal cut part. Reference these numbers on your cardboard template and cut sheet.

And to be even more organized, mark each part with a code number. The same code number should go on both the cardboard and the real material version of the part. This way you know what matches to what.

When you create your cut sheet, trace the cardboard parts onto the sheet, and write the same code number right on the cut sheet within that part's outline. If a hardware inspector asks you where on your cut sheet a certain part appears, simply look at the real part on the robot for the code number, then find it on the sheet.

 # 67. TRICK: Design for easy repair. Consider drilling "access holes" to allow easier disassembly.

During competition, you often face small technical problems and unanticipated repairs. To make life easier on yourself, make sure ahead of time that every part is accessible and easy to fix. Time is often limited between rounds and you need to make any necessary pit repairs as quick and pain-free as possible. If any specific part is hard to reach or hidden by another mechanism, you should consider drilling access holes. These are small holes in non-crucial parts that make it easier for you to dissemble or make repairs.

A prime example is the little black screw that holds a servo horn onto the servo. If you, for example, attach a flat bar across the servo hub, you would be well advised to drill an access hole in the bar that is large enough to accommodate the head of the screw that holds on the servo horn. This allows you to tighten the screw or remove the horn easily without first removing the bar assembly from the horn. That would make swapping a burned out servo motor far simpler—you would only need to remove the small center screw to detach both the bar and horn, rather than first remove two screws holding the bar on the horn then the servo horn screw.

 # 68. TIP: When assembling a new structure, put all screws in place with nuts fastened only loosely and then tighten them once they are all set.

If you put in a single screw and immediately tighten it before inserting the others, you may find that it is difficult to get the rest of the screws in place because there is no "wiggle room" to align the holes. This is especially true when assembling servo motors in their brackets, but also may come up any time a structure is held by more than a single screw.

 # 69. TIP: Protect wheels if possible. Leave room for wheel guards.

Leaving about an inch and a half on each side of the robot will provide enough room for a bent strip of metal to be placed on the side of, but not touching, the sides of the wheels. This greatly reduces the chance of a wheel falling off from both being loose and from a collision with a robot or game object.

70. *TIP: Protect chains with plastic or metal chain guards where practical.*

Using a chain guard nearly eliminates the risk of both entanglement with another robot via the chain and chain sprockets not securely fastened from sliding off their axles. We learned this at a scrimmage when in one round our partner robot slightly collided with our robot, causing our chain to jump off the sprocket, disabling our harvester. We added a chain guard and that never happened again.

71. *TRICK: Design the initial chassis to be at most 14" to 16" wide and long. This allows room to attach effectors and mechanisms that stick out past the wheel base.*

Thinking and planning this aspect beforehand reduces the stress of redoing the chassis to fit the requirements of the end effectors and saves a lot of time in the long run. As with many of the tips in this book, there are of course exceptions based on specific game details. But we have found that if you design the base chassis to be exactly 18" wide or long, you are not leaving yourself any room to attach certain kinds of effectors, and you might later find you must completely reconstruct the robot to give yourself more room, wasting time.

72. *TRICK: Build for rough play. Consider the outcome of different types of collisions and design guards and bumpers to minimize impact.*

During competitions robots go through numerous collisions and can often be damaged. Because of this it is a good idea to make your robot as durable as possible to avoid repairs, and the need for replacement parts. If possible make wheel guards, or covers, and mount sensors as firmly as possible. Mount servos or actuation mechanisms as far inboard as possible so these sensitive parts aren't damaged in a collision.

73. *TIP: Design so that wires are protected; position motors so that wire end is inside the robot body.*

Many students underestimate the stress and impacts the robot will need to survive in competition, including collisions with your own partner robot or field elements. Try to position wires so they cannot easily become entangled by such collisions. When given a choice, make sure the end of motors that contains the wire connections is facing inside the body of the robot as opposed to being exposed to collisions.

74. *TIP: Take the protective plastic backing off sheet plastic or sheet metal before drilling. The heat of drilling can melt it, making it hard to remove later.*

We learned this one the hard way. Keep the backing on for sawing to reduce scratches, but not for drilling.

Tools and Workshop

 # 75. STRATEGY: You don't need lots of expensive tools to be a championship team. Build your workshop tools over time starting with the most important items.

Do you think every championship team must have a CNC mill machine at their disposal, or a machine shop full of fancy equipment? Guess again! All you really must have is an electric hand drill with assorted bits, a hacksaw, an assortment of hex keys, some metal files, tin snips for cutting plastic shapes, a small vise clamp, some screw drivers, and several sizes of pliers to get started. For a hundred bucks worth of tools you can build a competitive robot.

If you have a little more budget then a bench top drill press and perhaps a bench top grinder would be our next suggestions. These make it easier to drill holes precisely and to grind cut metal neatly to avoid sharp edges or make required shapes. Bench top tools like this can be obtained for well under $100 each. A hand held jigsaw can be useful for cutting plastic or PVC. With further budget, the next item would be a bench top band saw (about $100 to $200). It is useful for cutting metal and plastic nice and straight or making gentle curves. Always follow all manufacturer safety rules for power tools, and adult supervision and proper safety and use instruction to students before using such tools is a must.

Finally, if you have all of the items mentioned above, and there is additional budget, you can think about adding a CNC router or a 3D printer. These items will allow you to develop 2D and 3D modeling skills and directly fabricate parts from your models. Parts made by these machines can be very accurate and look highly professional. Better yet, you can create duplicate parts that are exactly the same as the original versions. Many schools are now acquiring 3D printers or CNC routers, which are becoming increasingly important in real world engineering, so the time spent learning how to use these advanced tools is well worth the effort.

 # 76. TRICK: It's useful to have a #6-32 tap so you can thread holes.

A tap allows you to create threaded holes that accept the #6-32 screws allowed in the FTC competitions. You simply drill a hole slightly smaller than #6-32 (recommended drill bit size: index #32 or 7/64") and then insert the tap and turn it slowly to cut the threads in place. This allows for a neater and more compact fastening job in some cases than using a nut on the other side. (Certainly you will mostly use nuts; this technique is useful in certain cases when space is tight or a neater appearance is desired.) You can buy a #6-32 tap inexpensively from www.microfasteners.com.

77. TIP: *T-handle hex keys in the Tetrix sizes, a T-handle 5/16" nut driver, and 5/16" open and closed end combo wrenches will save you lots of time.*

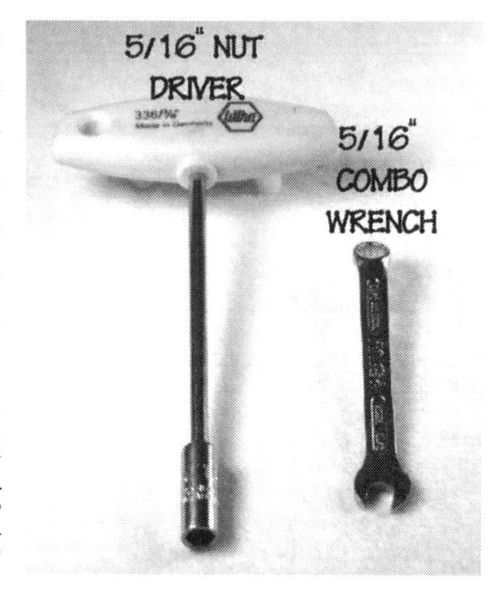

If you just use the simple hex keys that come with the Tetrix sets you will waste a lot of time and experience endless frustration. You can buy high quality T-handle hex keys and nut drivers for less than $7 each and they will pay for themselves many times over in terms of saved time and eliminated frustration. An open/closed end combo wrench also lets you reach nuts in tight spots and is frequently useful.

This is not to say you will never use the small hex keys in the Tetrix set. Sometimes they are quite useful for reaching tight spots, such as when tightening wheel hubs (the wheel rim overhangs the hub making it hard to reach with larger tools).

78. TIP: *A great place to buy hand tools is www.wihatools.com (Wiha Premium Tools).*

They have T-handled hex keys, nut drivers, nice screw drivers, and open and closed end wrenches in all the sizes you need. You can purchase just the sizes required by the Tetrix kit. The tools are high quality and not terribly expensive. With reasonable care they will last you many seasons.

79. TIP: *A great place to buy all kinds of tools is www.harborfreighttools.com (Harbor Freight Tools).*

Many hobbyists use Harbor Freight Tools for both hand tools and electric tools. Prices are low and service is great.

80. TIP: *You can find some interesting low cost surplus tools and other useful items at www.sciplus.com (American Science & Surplus).*

American Science & Surplus has items ranging from the practical to the totally bizarre. They specialize in low cost surplus items. You can not only find great deals on tools, but odd items that could make eye-catching decorations for your pit, give-aways for tournaments, or even team uniform accessories.

 # 81. TRICK: A carpenter square plus a T-square allows you to easily make sure your robot does not violate size restrictions.

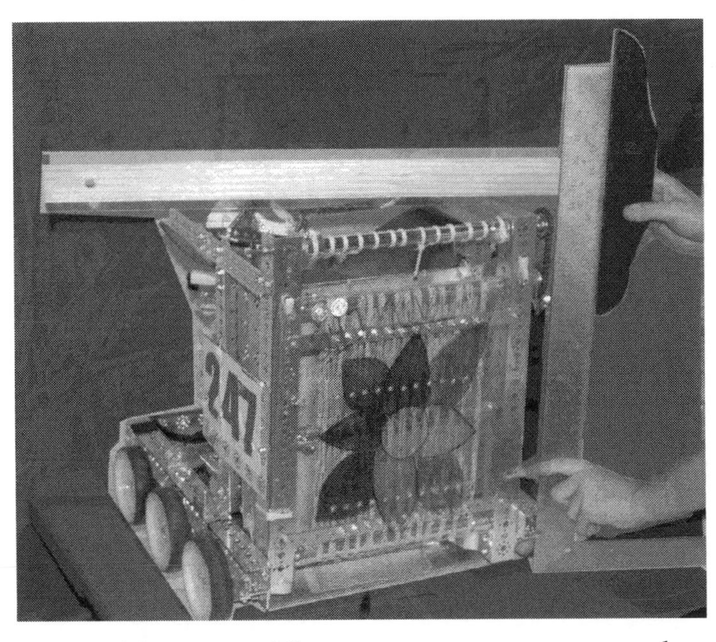

You'd be surprised how hard it is when you're first building a prototype robot to make sure you're not going outside the 18" cube size restriction, or to determine exactly how much space remains. Of course you could build an 18" cube box, and you probably should do that for a final check, but even that doesn't really help you to know precisely how much room you have remaining to bolt on your latest and greatest effector.

A simple solution is to obtain an L-shaped carpenter's square as well as a T-square. The carpenter square can be placed alongside the robot to easily measure height at one side, then line up the T-square at the 18" mark on the carpenter square and you can see at each point how tall your robot is. A similar procedure can measure the width or length of the robot easily.

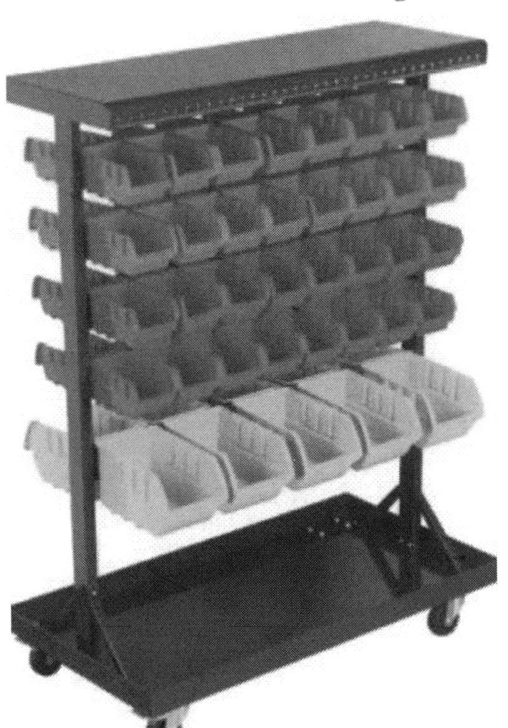

But do remember to check your size very carefully before every competition. Nothing is more hectic than getting rejected during hardware inspection because you are a quarter inch too long, then spending time frantically hack sawing bits off your robot before your first round of play!

 # 82. TIP: A great way to organize the (many) robot parts is a double-sided, open bin mobile rack.

As of this edition, this item is sold for about $100 from Harbor Freight Tools, part number 95551. It has 74 bins, which is enough to store all the different Tetrix structural parts plus electrical parts, sensors, motors, etc. Small parts like screws, nuts, washers, spacers, and axle set screw collars should probably be kept in tackle or bead boxes, but larger parts work best in open bins.

We have found this is superior to packing the parts in closable boxes, because such boxes take up a lot of table top space and it is quite difficult to locate parts that tend to get all jumbled up in a closed box system. Open bins make it easier to find the part you're looking for.

Materials and Fasteners

 # 83. TIP: Use nylon insert nuts.

Sometimes called "Nyloc," they have far, far more holding power than the KEP nuts sold on the FTC parts site. They are harder to tighten and loosen, but if you have a good set of nut drivers, T-handle hex keys, and open and closed end wrenches you'll have no problem, and the time you save by not constantly having to tighten all the loose screws is worth a little extra effort putting them on the first time. Please note: in recent seasons nylon insert nuts have been allowed as a fastener, but please check the game rules and forum postings to be sure before using them in any new season.

 # 84. TIP: A great place to buy bulk quantities of #6-32 screws, washers, nylon insert nuts, and related tools is www.microfasteners.com (Micro Fasteners).

You can purchase both button head and cap head screws there, as well as assorted washers, nylon and standard nuts, and even some nice screwdriver-handled hex keys and a #6-32 tap in case you need to tap your own threads. The variety of screw sizes available is much wider than available on the Tetrix order form page, and the prices are reasonable, with quick shipping by Priority Mail. Standard plated screws are $4 to $6 per hundred for sizes under an inch long, more for longer sizes.

 # 85. TIP: Keep your screws, nuts, and washers sorted using a plastic bait or bead box.

It's a lot faster finding the exact size screw you need by looking through a sorted set of bins in a single box than hunting through a big pile on the table. Plastic bait and bead boxes are inexpensive, available online or at many stores like Wal-Mart, Target, or sporting goods stores, and frequently have adjustable compartments to accommodate both long and short screws nicely. Keep button head screws separated from cap head screws, and organize sizes to increase from one side of the box to the other.

 # 86. TIP: At cleanup time, sweep loose parts and screws into a box; students who have "down time" at the next meeting can sort them into the bait box properly.

Taking a little extra time during cleanup to sort loose screws on your workbench back into their proper places will yield saved time later. Also, making students sort out screws when they aren't doing anything encourages them to keep busy!

 # 87. TIP: As parts are removed from the robot or from early prototypes, check them for damage. Discard or mark parts that are no longer perfect.

Bolts get stripped, bronze bushings get worn out, and axles bend. If a part is damaged, throw it away or perhaps mark it and put it in a special box of "worn parts" that maybe are only used for prototyping or non-critical functions. You don't want to use already damaged parts on your competition robot by accident.

 # 88. TIP: A great place to buy allowed metals, plastics, nonslip pad, surgical tubing, and other materials is www.mcmaster.com (McMaster-Carr).

Even the FIRST game manual in some years has mentioned McMaster-Carr as a source of permitted materials. Just make absolutely certain the material sizes and thicknesses you order conform to the limits set out in the new season's game manual.

 # 89. TIP: A great place to buy plastic and rubber materials is www.usplastic.com (US Plastic Corp.)

The new season's game manual frequently allows various kinds of plastic sheet, just make sure you adhere to the correct type of plastic and the maximum allowed thickness and dimensions outlined in the game manual.

 # 90. TIP: A great place to buy sheet aluminum and plastic sheet is www.onlinemetals.com (Online Metals).

There are different grades of aluminum and some grades are far more expensive than others, so be sure to check. We've found the very cheapest grade works just fine. Online Metals will actually cut a specific size rectangle for you, which could save you money if you don't need the full size allowed by the game rules (in past years typically aluminum sheet 1/16" thick, 2 feet square has been allowed.)

 # 91. TRICK: In general, do not use any LEGO structural parts.

Other than perhaps decorations, you are almost always better off using metal (or sheet plastic) rather than LEGO structural parts like bricks or axles. Note that this tip only applies to *structural* parts and excludes LEGO electrical parts, motors or sensors, which can be critical to success. We're talking about using plastic bricks, etc. for your robot's structure.

You will rarely see a LEGO structural part on a Championship robot for any critical function. Frequently, when you do see teams use LEGO parts for structures like sweepers or harvesters, the result is that by the end of a round of play little LEGO bits are all over the field! They just don't stand up to the rigors of competition. There can be exceptions (low-stress mechanisms like agitators) and if properly reinforced with zip ties and cross beams it may be possible to use them for some purposes. But when in doubt, don't use LEGO plastic parts for structural elements.

Chassis and Structural

 # 92. TRICK: Tie together frame elements using short channel rather than an L-Bracket.

If you use an L-bracket to tie together channels into a frame, you will have an annoying bit of space between the channels. This weakens your structure. One of the very short pieces of channel makes a much better connection.

In the picture, a short piece of channel with one side cut off makes a perfect connection between two long channels.

 # 93. TRICK: Another way to tie together channels is to use a piece of flat metal.

Cutting a piece of the Tetrix flat metal plate allows you to strongly and snugly connect frame elements, especially to form a rectangle for your base chassis, or, as pictured here, to hold two parallel channels together.

 # 94. STRATEGY: Make the robot heavy.

Although we cannot guarantee this will be true for all future games, past games have often favored heavy robots. These are harder to push around when jockeying for a position and have more momentum when it becomes necessary to ram opponents who are playing defense. A heavy robot also gets much better traction on the field surface. We have seen some teams go so far as to attach "ballast" to their robots, including extra gears that do nothing but add weight and even unused motors. (Be aware that you are still restricted to a set number of motors whether they are actively being used or are just there for weight.)

There have been recent exceptions to this rule. The 2012/13 season game "Ring it Up!" featured an endgame bonus for lifting your partner robot. Obviously it's easier to lift a lighter robot. It turned out not to be a major factor in the game, however, as it was often more effective to simply continue scoring in other ways during endgame. The 2013/14 game "Block Party" features an endgame bonus for hanging your robot off a bar. Again, the heavier your robot is, the harder it's going to be to pull it up onto a bar. So there are limits to how heavy you can make the robot and still score this way in the endgame.

 # 95. STRATEGY: Unless there are specific game factors that favor small robots, make the robot large.

Again, it is possible that future games will have rules that favor small robots, but our experience is that you almost never see a small robot win a championship. The robot should be near the maximum allowed size if possible. The reason for this is that a larger robot can more effectively occupy strategically crucial space, denying that space to opponents. A large robot, even if it is not effective at scoring, is more likely to be chosen as an alliance partner for defensive purposes than a small robot that can be easily pushed around. Your chances of making it to the elimination rounds are usually improved if your robot is big and heavy.

One quick update for the 2010/11 season: the game "Get Over It!" required the ability to balance on a bridge while carrying a rolling goal, preferably along with a partner robot doing the same. A smaller robot could have some advantage in this game because there would physically be more room to fit more items on the balance bridge. This game also was relatively "crowded" with many field elements taking up a lot of floor space compared to past games. Robots that were too large could find it hard to maneuver on the crowded field at times.

Even with all these factors, many robots that made it into the elimination rounds were fairly close to the legal size limits, but there were cases of smaller robots gaining some advantages due to specific game factors.

The upshot is: keep an open mind and really analyze the new season game to see if smaller sizing might have an advantage or not.

 # 96. STRATEGY: Keep the center of gravity as low as possible.

Try to mount the battery, motors, and other heavy elements near the bottom of the robot. Favor designs that keep heavy structural elements low. Most robots will have trouble righting themselves after tipping over (or perhaps can only right themselves if they tip in a certain way.) At the very least you will waste time trying to get back onto your drive wheels after tipping over. A low center of gravity

makes tipping far less likely. This is especially important in games that involve ramps such as the 2008/09 season game "Face Off."

But even without ramps, robots do tip over while wrestling for positions on the field, and most of the cases we've seen involve robots that have violated this tip, for example mounting their battery on top of the robot or having an actuator that rises up and makes the robot very tall. And unlike the "large" and "heavy" tips above, it's hard to imagine any game that will ever favor a robot that has a high center of gravity!

 # 97. TIP: Design to prevent game elements from getting stuck under your robot.

This is a frequent problem in games. In the 2009/10 season game, "Hot Shot!" it was quite easy for that year's scoring element (a wiffle ball) to become trapped under the robot's chassis, making driving difficult or impossible.

Typically you will want to design your robot so it has metal or plastic guards all around to prevent game elements from becoming trapped under the robot chassis. In the picture, a piece of channel hangs between the wheels to prevent the game element from rolling under the chassis.

The Drive Train: Gears, Sprockets, Chains, and Wheels

 # 98. TIP: Yes, there is a difference between a motor shaft hub and an axle hub!

There are two kinds of "hub" used to connect wheels or gears to axles or motor shafts. They look almost identical, but it actually turns out to be quite important which one you use. The motor shaft hub has a very slightly larger hole in the middle to admit the shaft of the 12v motor. Realizing there is a difference between the two is the first step to avoiding frustration. If you try to use the motor shaft hub on an axle, it will be too loose and you'll get a wobbly gear or wheel. If you try using the axle hub on a motor shaft, it won't fit. We've seen students trying to bang an axle hub onto a motor shaft with a hammer, in vain, and probably damaging the motor to boot! Learn the difference!

 # 99. TIP: Keep drive wheels symmetrical.

While this may seem obvious at first glance, we have seen many robots where a wheel on the left side was mounted in a slightly different position than the corresponding wheel on the right side. This causes turns in one direction behave differently than the other direction, making it harder for the driver team and the autonomous programmer to get consistent behavior.

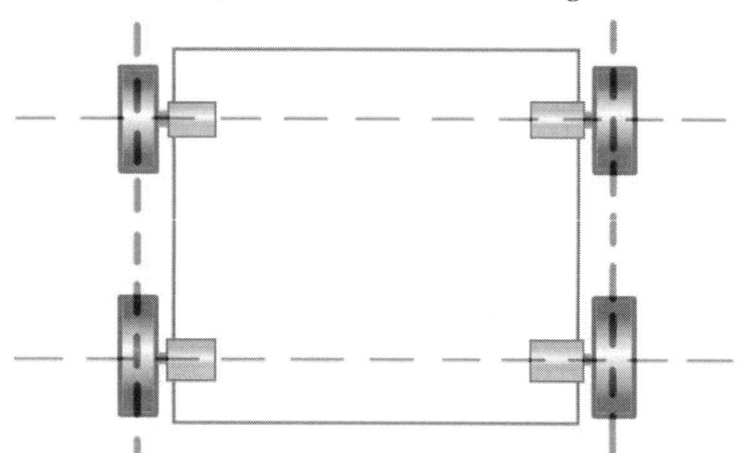

 # 100. STRATEGY: Direct drive wheels are simpler, easier to keep aligned, and easier to repair, if the power available meets your game strategy needs.

Direct drive means the wheels are directly connected to the motor shaft with no gears or chains involved. Whether you can use direct drive depends on the details of the game. Basically, we favor direct drive whenever the game does not call for high speed or high pushing power, because there are many advantages to direct drive. Gears can easily become unmeshed and grind or lock up. Chains can get snagged or come loose. Direct drive motors are easy to swap if they burn out, gears or chains take a lot more coercing.

As usual, there could be good reasons to ignore this strategy and use gears or chains to drive the wheels, depending on the game's specific details. If you do go with gears or chains, you simply need to carefully check gear alignment and chain tension between each and every round of play to keep them operating in tip top shape.

But for games where direct drive wheels offer sufficient speed and power, the advantages are compelling in terms of reliability and simplicity.

 # 101. STRATEGY: Geared wheels are less likely to damage the motor in a collision.

Yes, we realize this tip contradicts the one above! Over and over again we find that engineering involves trade-offs. There are advantages to direct drive wheels, and there are other advantages to geared wheels. Because the force of collisions does not bear directly on the motor shaft, geared wheels offer some protection for the motor.

 # 102. TIP: Know how your available gear ratio choices affect speed and pushing power.

The gear ratio and wheel size determine speed versus pushing power. Generally speaking, gearing up increases speed and decreases pushing power, and gearing down does the opposite. The table below shows how gear ratio and wheel size affect speed and maximum possible pushing power. (Pushing power may be less than what is shown if you can't get sufficient floor traction, however, so see the discussion in the next TIP.)

Gear Ratio	Wheel Size	Max. Speed (feet per sec.)	Pushing Power (oz. per motor)	Notes
Four Inch Wheels				
1:3	4"	0.88	450	
1:2	4"	1.33	300	
2:3	4"	1.77	225	
1:1 (direct)	4"	2.65	150	
3:2	4"	3.98	100	
2:1	4"	5.3	75	Not recommended
3:1	4"	7.95	50	Not recommended
Three Inch Wheels				
1:3	3"	0.66	600	
1:2	3"	0.99	400	
2:3	3"	1.31	303	
1:1 (direct)	3"	1.99	200	
3:2	3"	3.98	100	
2:1	3"	5.3	75	Not recommended
3:1	3"	7.95	50	Not recommended

Note that this table is constructed by just using the motor specifications and doing straight math on the gearing and wheel diameters, it does not take into account friction or traction effects. Calculated this way, it doesn't matter whether you use gears or chains (although there is a 3:2 ratio available with sprockets). But in practice, there are differences between gears and chains even at the same ratios both in terms of how they are used to build the robot and different kinds of problems that may occur with one or the other.

In reality, gearing will reduce the power shown in this chart because of friction; the higher the gear ratio the more such friction losses can be expected to occur. Therefore, the direct drive figures in this table are probably the most accurate, while the higher gear ratios will not attain the full specifications listed here due to these friction losses.

Pushing power listed in this table is also per motor. If you have four motors driving the wheels then multiply the figure by four, for example.

Note that there is a fairly large difference between using the 4" wheels vs. the 3" wheels. For example, the 3" wheels using direct drive give you reasonably similar speed and power to the 4" wheels at a 2:3 ratio, but with the benefits of the simplicity that direct drive offers. But there are other considerations: smaller wheels give your robot less ground clearance for example, which may or may not be important to your game strategy. As we see over and over again, engineering is all about choosing from different tradeoffs.

 # 103. TIP: To maximize pushing power you need enough traction to avoid wheel slippage.

In addition to the discussion in the prior TIP, the "pushing power" listed in the table can only be attained if there is sufficient traction on the field surface to prevent the wheels from slipping. Traction is a function both of the weight of your robot (the more weight the more traction), the number of wheels (the more wheels the *less* traction unless the surface is uneven allowing only certain wheels to grip, because more wheels distribute the weight of your robot over more surface area), and the gripping power of the wheel itself against the floor surface.

Some teams enhance the gripping power of their wheels by doing things like scoring the rubber surface with a hacksaw or even looping cable ties around through holes they drill around the radius (similar to chains used on cars in the winter). However, such enhancements may decrease turning accuracy during autonomous so there is, as usual, a tradeoff. Sometimes, too much traction can actually make the robot harder to drive and turn.

 # 104. TIP: Designs that use a motor for each and every drive wheel are more robust against failure.

We also favor one motor on each and every drive wheel when possible, whether geared or direct drive. This way if a motor gets burned out you might still limp along with three of four drive wheels operating. You also obtain more pushing power without gearing down. We have won matches even after losing a drive motor this way. But if you only have one motor on each side of the robot, losing one of those motors leaves a robot that can only spin around uselessly in circles. Championship teams try to design their robots to avoid single points of failure in major subsystems, giving themselves a chance to win even when hardware fails.

Once again, the details of the game and specific rules may force you to violate this tip. For example in the 2008/09 season game, "Face Off," the rules only allowed a maximum of four 12v drive motors and most teams' designs required at least one 12v motor for an actuator.

 # 105. TRICK: Near motors and gears, re-enforce channels with metal offsets to keep them rigid.

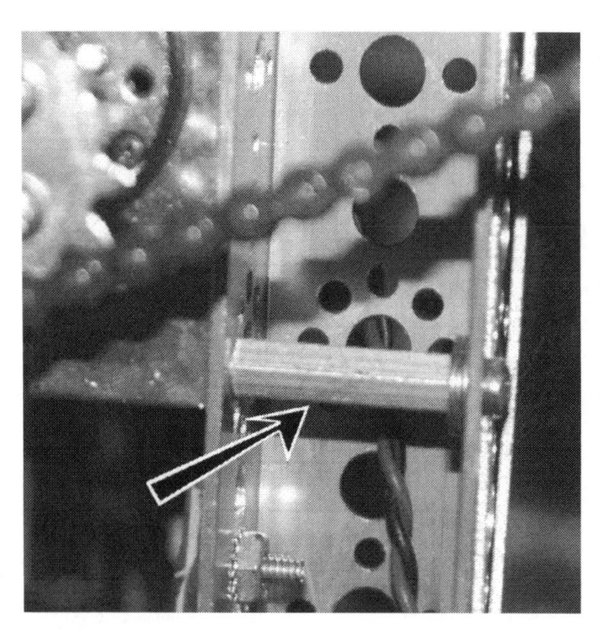

If you don't do this then the channel can warp under stress, causing gears to grind or even freeze up, or chains to strain. This can lead to motor burn out and a catastrophic failure of the drive train or actuator. This is especially important when gearing up or gearing down, or for actuators that will have to lift heavy loads. Do not assume that the channel will remain rigid without reinforcement!

In the picture, a short metal offset and two washers are used to reinforce the channel under a motor mount that drives a chain. This keeps the chain from sagging under load.

 # 106. TRICK: A chain breaker is easier to use than the chain link tool sold on the parts site.

There are relatively inexpensive (under $20) chain breakers that work much better than the one sold on the FTC parts site. An example of a low cost chain breaker can be found at www.thebigbearingstore.com (The Big Bearing Store) as well as in many other places. Make sure you get the size for a #25 chain. These tools make it far easier to align the tiny pin and push them out, and have an easy to turn T-handle built right in. Online tutorials that show you the proper way to use a chain breaker can be found by googling the phrase: *using a chain breaker.*

 # 107. TRICK: Use half-links (also called offset links) to make a chain slightly tighter.

When you connect a chain you have cut to size, sometimes you find it is still slightly too loose. You can use a half-link connector instead of the full-link connectors that come with the chain sold on the FTC parts site. Sometimes these are called offset links. You can purchase them from stores that carry the #25 roller chain allowed by the competition, such as at www.thebigbearingstore.com (The Big Bearing Store) and others. The FTC forums in the 2009/10 season explicitly confirmed that half-links were allowed, but please check the game rules for the new season to confirm that has not changed before using this trick.

 # 108. TIP: A chain is the correct tension when the center of the looser span moves about 3-5% of the center-to-center distance between sprockets.

There is a tighter span and a looser span of chain between the sprockets (which side is tighter depends on the direction of motion and other factors). The center of the looser span should be able to move approximately five percent of the distance between the centers of the sprockets. If too tight, the chain is under too much stress which will cause wear on your motor and may even snap. If too loose, the chain can easily pop off.

So for example, if the distance between sprocket centers was 8 inches, there should about 0.4 inches of movement possible in the center of the looser span of chain.(5% of 8 is 0.4)

See the photograph, in which the mid-span movement is illustrated using a "ghost" image of how high the chain can be pushed vs. how low it can be pulled using modest finger force. In this case, the mid-span movement is about fifteen percent of the sprocket center distance, so the chain is too loose.

 # 109. TIP: You may need a "take up" mechanism if you have more than two sprockets on a single length of chain.

Going beyond two sprockets on a single length of chain complicates things. Depending on exactly how things are arranged, you may need some kind of take-up mechanism that ensures the chain does not go slack around one or more of the sprockets. This could be as simple as some nylon offsets on a screw used as a roller, or as complex as a flexible mechanism that uses rubber bands or surgical tube as springs to keep tension correct on the chain.

When possible, of course, it might be simpler to favor designs that do not require more than two sprockets on a chain, avoiding this situation entirely.

 # 110. TIP: Avoid having a larger gear on the motor shaft driving a smaller gear on the drive wheel at more than a 3:2 ratio.

For example, don't put an 80 tooth gear on the motor shaft driving a 40 tooth gear on the drive wheel, because that would be a 2:1 ratio which is higher than a 3:2 ratio. Violating this tip frequently leads to grinding gears and smoked motors. In our experience you can get away with a 3:2 gear ratio but you're asking for trouble if you go to a 2:1 or 3:1 ratio. You're better off using chain if you want to "gear up." Another technique is to have the motor drive a second gear at a 1:1 ratio then have that second gear drive the wheel or actuator gear at a higher ratio.

 # 111. TIP: Gears are meshed correctly if you can just barely fit a piece of paper between the teeth.

Gears that are too tight will freeze up and smoke your motor. Gears that are too loose will slip and grind against each other, which frequently chips and dulls the teeth of the gears. Gear spacing can be adjusted by twisting the motor in the motor mount, because the shaft is offset from the center. So you should initially mount the gears such that the motor shaft is about halfway between its maximum and minimum distance from the corresponding gear, then twist until a piece of paper just fits between the teeth and can be removed without ripping (standard printer paper, some people use a dollar bill). Then you lock down the motor mount tightly. Check the spacing between rounds of competition and adjust if necessary.

 # 112. TRICK: Gears should line up edgewise perfectly in order to avoid grinding, and should be locked down with spacers so they cannot move along the shaft or axle.

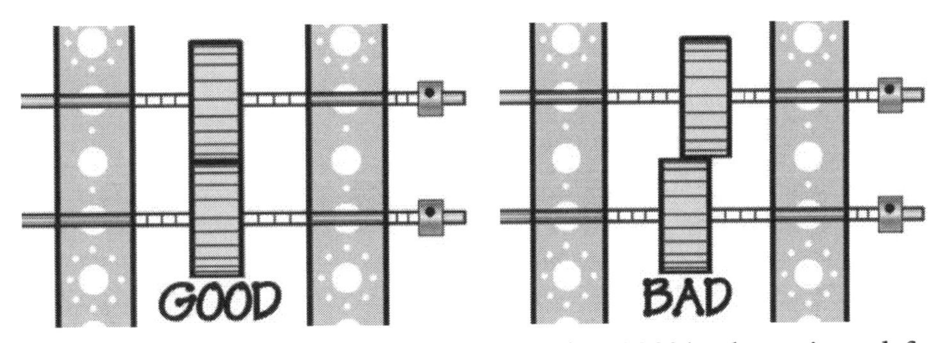

You must use spacers and other techniques to make sure gears are spaced properly both tooth-to-tooth as well as along the axle side-to-side. The gears must overlap 100% when viewed from the edge (see diagram). Lock down everything with spacers; do not depend on a shaft position staying put just based on the set screws. One or two collisions and your gear may "walk" down the shaft and become fully or partially unmeshed with its mating gear. That grinding sound you then hear is your chances of victory slipping away…

 # 113. SECRET: Chain can be used to create a conveyor belt by cutting "bypass" holes in sprockets and slipping screws through the links to hold conveyer materials.

It is possible to slip standard #6 screws in between the links in the #25 chain, creating an incredibly sturdy conveyor belt system.

The screw needs to be sanded down lengthwise using a belt sander until about half of it is gone. The screw will then slip into the large links, and can be secured with a nut.

Bypass holes will need to be sanded into the sprockets driving the chain and the screws and bypass holes must align so that they mesh at exactly the right times. It is difficult but possible.

In the photograph, the head of a screw holding a conveyor

mechanism to the chain exactly meshes with a bypass hole ground into the sprocket. This also means that all of your screws and conveying materials must be exactly evenly spaced.

Note: It is important to use nylon insert nuts for this job. The screw threads will be weaker because they are partially ground away, so a sturdier nut is necessary.

 # 114. SECRET: Six wheel designs should use washers or other techniques to make the center wheels on each side slightly closer to the ground than the outer wheels.

This allows for accurate turning as only two of the three wheels on each side has total grip on the floor. If you don't do this the wheels will resist accurate turns which would be bad for autonomous mode. If you use direct drive on all six wheels then you can adjust the motor's off-center axle to achieve this, otherwise you can use a couple of washers. It only has to be a sixteenth of an inch or so difference.

 # 115. SECRET: For gear or chain driven mechanisms that may jam against game elements, consider building a torque limiter to reduce stress on components.

A torque limiter is a device that keeps mechanisms safe by limiting the amount of force the mechanism can apply. Once the maximum force occurs, the torque limiter is designed to "slip" so as not to damage motors, gears, chains, etc. An easy way to construct a torque limiting device with the parts available in FTC is a friction plate type mechanism. A nylon servo horn inserted between two hubs, with adjustable screws to provide a tunable amount of friction works well. When the effector attached to the limiter is overloaded, rather than smoking the motor, the limiter slips, preventing serious damage.

In the picture, two nylon servo horns (see

arrows) are sandwiched around the large gear, which has no other connection to the shaft. The screws sticking out on the left side of the photo can be adjusted to squeeze the servo horns either tighter or looser against the big gear. The big gear will slip if the force it exerts exceeds the friction of the servo horns. By purposely slipping once a dangerous amount of force occurs, the torque limiter saves the motor and mechanisms from excessive stress. Motor burnout is a major cause of systems failures that result in losing rounds of play.

Pivots and Actuators

 # 116. TIP: For low stress pivots, a screw can work just fine. Double up the nut so it can remain loose enough to allow motion.

If a pivot point only needs to bear a small amount of weight, keep it simple and just use a screw as the pivot. Make sure you lubricate it and use washers. Keep it slightly loose by doubling up the nut on one side, tighten the nuts against each other so the pivot remains mobile.

Important note: in recent years the forums have ruled that if you use a screw other than the ones available on the Tetrix parts site then the only allowed use of such a screw is as a "fastener." This implies that to be totally safe on materials usage you should only use official Tetrix screws as pivots, since a pivot usage is not consistent with being a "fastener." Please carefully check the new season rules and forum postings to be sure you usage is allowed.

 # 117. TIP: For pivots that need to bear weight, use the bronze bushings.

The bronze bushings are impregnated with oil and make a great pivot. They are just large enough to admit an axle. If necessary, drill a hole just a tiny bit larger than the bushing in whatever metal structure will hold the pivot. The Tetrix channels have bushing-sized holes pre-drilled, but not the flat bar or angles.

 # 118. TIP: Try not to attach screws too close to bronze bushings, because the nut may compress the bushing and cause friction.

If you must place a screw near a bushing, make sure the nut is turned so that the flat face of the hexagon is turned toward the bushing, not a sharp angle that could dig in. But when possible try to avoid this problem entirely by locating screws elsewhere.

 # 119. TIP: Learn how to properly use the single and double servo brackets with associated pivot bearings by viewing the tutorial at the TetrixRobotics.com web site.

The tutorial shows proper mounting of both single and double servo pivots. Proper mounting will mean less stress on your servos and less chance of failure. As of this writing the tutorial is at this URL:

http://www.tetrixrobotics.com/Building_System/Servos_and_Pivots/

 # 120. TIP: Rubber bands are a reasonable substitute for light duty springs. Surgical tube, if allowed by the season's rules, can handle significant spring loads.

As of the 2010/11 season it is not legal to purchase commercial springs for use in an FTC competition robot. Although in theory you could construct your own spring this is not really practical given the materials allowed. A reasonable substitute for parts that need some kind of spring action are rubber bands, and for heavier loads quarter inch surgical tube, both of which have been allowed recently. However, make sure the game rules still allow them, as the rules do change from year to year.

In addition, be aware that surgical tubing, if stretched too far, can store a very substantial amount of kinetic energy and could represent a danger. Only use enough tension to solve the problem, and take care when using it.

Motors, Servos, and Controllers

 # 121. TRICK: Mount all your motor and servo controllers on the bottom of the robot, facing down so you can get at all the connection terminals.

This is a great place to mount the controllers. You can get to them easily by flipping the robot on its side. They are protected there, and they keep your center of gravity low. As you add actuators and other mechanisms during the season, you never have to worry about having trouble getting to your controllers. If you want to, you can even mount a piece of clear plastic over the controllers for further protection in case a game element gets caught under your robot.

 # 122. TRICK: Servos are not very strong but can lift significant weight if "ganged" together and rubber bands are used to provide "counter weight."

Pictured is "Waffles;" one of our robots from the 2008/09 season. It uses six servos "ganged" together to lift a bucket used to score game elements (see top right side of picture). Even the six servos were not enough to lift a bucket full of pucks, so note the rubber bands running diagonally from upper right to lower left. These serve as a counter-weight to take some of the stress off the servos. Most teams that season used a 12v drive motor for their scoring mechanism. Only four 12v motors were allowed that year. By using servos instead, we were able to place all four 12v motors on our drive train, allowing for both speed and pushing power, which turned out to be a major advantage.

Mounting Batteries and the NXT Brick

 # 123. TIP: For safety, make sure there is no screw or nut or any sharp object under your battery's mounting position. Only flat metal or plastic should be under (or around) your battery.

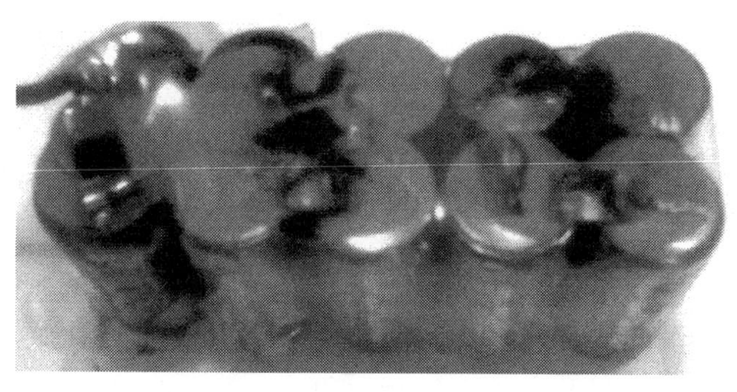

We found this out the hard way. Less than a quarter inch of the end of a screw was sticking up under our battery compartment. After many rounds of play, the screw worked its way through the insulation. The battery actually caught fire during the 2008/09 Monty Madness post season event (see picture of our scorched battery).Fortunately, the refs were able to put it out, nobody was hurt, and despite the robot catching fire we still won that round!

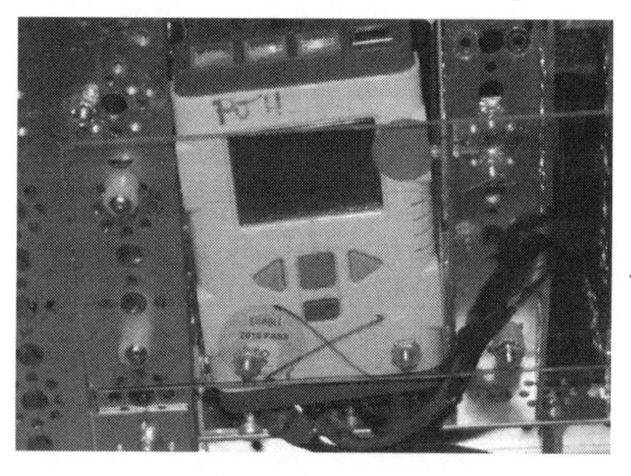

 # 124. TRICK: Mount a piece of clear plastic over the front of the NXT Brick to hold it in place and prevent accidental button presses, but do so in a way that you can slightly lift the brick to hit the buttons yourself.

The buttons are slightly recessed on the front of the NXT Brick so you don't have to worry about the plastic accidentally pushing them.

 # 125. TIP: Always make sure the NXT Brick USB port is accessible when it's mounted in the robot.

Uploading new programs and recovering Bluetooth connections will be much easier if you follow this tip.

 # 126. TIP: It's also nice if the battery charge plug is accessible when the brick is mounted to your robot.

You can then easily plug it in for a quick top-off between rounds of play without even removing it from your robot or turning it off. Do recall, however, that it is a good idea to occasionally reboot your NXT brick.

Mounting Sensors

 # 127. TIP: If you slice the end off a flat bracket or L-bracket, the holes that remain line up perfectly with the holes in the sensors, allowing you to securely fasten the sensor to metal using #6-32 screws.

The picture shows a standard L-bracket with the rounded end of one side removed by hacksaw then ground smooth. The holes line up almost perfectly with the holes in the plastic body of the sensors. In this case some white plastic spacers were used as well, but whether you need them depends on precisely how you need to mount the sensor on your robot.

Alternatives to Using Tetrix Structural Parts

In the 2012/13 season and the 2013/14 season extruded aluminum parts became legal building materials. Since that time, our team has almost completely stopped using Tetrix (or Matrix) channel in favor of using slotted extrusions made by several manufacturers.

In our opinion, the advantage of slotted extrusion building systems are that they are less expensive than Tetrix or Matrix channel, they are stronger, and they allow fine adjustment of positioning parts due to the fact that you can move a part anywhere along a "slot" that is in each face of the metal.

The picture above shows MicroRax brand slotted aluminum extrusion. Note how the L bracket is held into the metal by slots. A nut slides into the slot (you can't see the nuts, they are below the L-bracket inside the slot). You can see the big advantage here: instead of being limited by exact placement of holes in something like the Tetrix channel, you could move the upper right piece of the L bracket anywhere within the slot. This allows easy adjustment: just loosen the two screws, move the L-bracket to the correct position, and tighten again. Note that brackets and fittings for this type of extrusion are also generally legal building materials. We also use a CNC router to create our own custom brackets and fittings from time to time out of HDPE plastic. But even without a CNC router these systems end up being less expensive than Tetrix structural parts and we have found them to be superior in almost every way.

Another big advantage is that some brands of extrusion also have linear bearings you can purchase that fit their extrusion system perfectly. This can allow you to create mechanisms that have a sliding motion very easily and reliably. Linear bearings of practically any type became legal parts in the 2012/13 season and remain legal in the 2013/14 season. As always, if you are reading this book later than 2013/14 you must check your specific game rules before using any part.

 # 128. TIP: A good source of small (10 mm) slotted extrusion parts is MicroRax. (www.microrax.com).

MicroRax 10mm extrusions are great for small mechanisms that need to be lightweight. They use M3 screws which are commonly available. There are a variety of brackets and bearings, most of which are currently legal parts.

 # 129. TIP: A good source of larger (15mm, 20mm, 25mm and more) slotted extrusion is Misumi (us.misumi-ec.com in the USA).

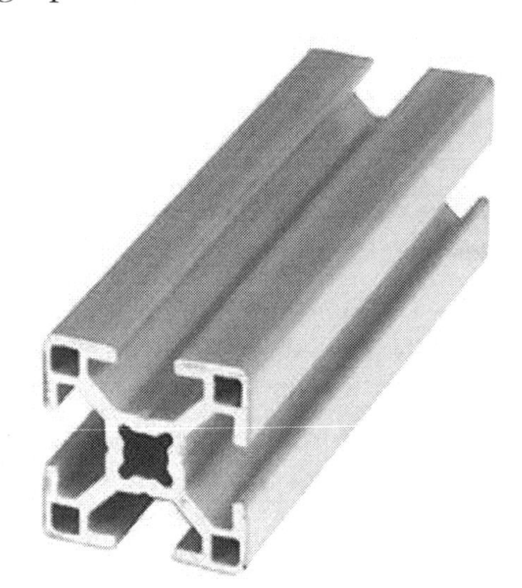

Misumi carries a very wide variety of extrusions from 15mm all the way up to 60mm. We find that for structures that need a great deal of strength and rigidity, generally speaking 25mm is sufficient. This would be what we use for a chassis or heavy lifting arm. The 15mm is great for most other purposes however. Misumi also carries many add-on parts that are currently considered legal such as brackets and linear bearings. They have a fantastic online search tool to find just what you're looking for. You can order long lengths and cut them yourself to size (which is very cost effective) or you can have them cut the lengths for you.

 # 130. TIP: A good source for larger (1", 1.5", 2", 20mm, 40mm, 45mm) extrusions compatible with those made by 80-20 corporation is McMaster-Carr. (www.mcmaster.com)

Once again McMaster-Carr is our one-stop shopping source for many different materials! The brand is called 80-20 but there are also compatible extrusion systems, and this type of metal has been used for many years by FRC teams. A very wide variety of brackets, linear bearings, and other currently legal extrusion accessories are also available there, all searchable with McMaster-Carr's excellent search tool.

 # 131. TIP: If you have access to a CNC router, DXF files are readily available that allow you to fabricate compatible plastic brackets for slotted extrusions.

We have found that 3/16 inch thick HDPE plastic makes terrific brackets for aluminum slotted extrusions. Better yet, you can download DXF files directly from the manufacturer or retailer's web site which can be used as the basis to fabricate your own custom parts. We usually make our own custom modifications to the DXF files first. For example, we made our own slotted extrusion motor mounts, allowing us to mount Tetrix motors directly to slotted extrusions. We made our own corner brackets that are perfect for typical FTC chassis designs, etc.

Many high schools now have CNC routers. They can be purchased for around $1,500 which is within budget for many teams. As with any big-ticket item, though, you are best served if you have access to a mentor or engineer who can teach you how to most effectively use the machine.

Note that you could in theory also use the DXF files to create models of brackets that can be 3D printed. We feel that under most circumstances the CNC router makes parts out of HDPE that are more rugged, however. 3D printed models have a "grain" to them where the layers are built up, and that grain is subject to stress cracking far more than a solid piece of HDPE plastic. See the following section on plastic for more details.

Plastic Fabrication and Thermoforming Tips

One area of fabrication that relatively few teams take advantage of is the thermoforming of custom plastic parts. In recent seasons the rules have allowed using various sizes and grades of plastic, and bending that plastic by heating it to its thermoforming temperature is a permitted fabrication technique.

As usual, we would caution you to check the new season's game rules to make sure plastic rules have not changed before you invest a great deal of time learning to thermoform. But the trend in recent years has been to allow more kinds and thicknesses of plastic to be used in the game. For example, in the 2010/11 season you could use four different kinds of plastic sheet (polycarbonate, Kydex, PETG, and ABS) up to $1/8^{th}$ inch thick, each sheet measuring 2 feet by 2 feet. That is a very large amount of plastic! In fact it would be possible to make a robot consisting of just a channel frame on which the drive train is mounted with almost the entire remaining structure composed of plastic!

Advantages of Plastic

Plastic of the types allowed in recent years can be quite strong as well as light weight, is frequently less expensive than metal, and has the added benefit of not significantly blocking WiFi signals from the Samantha module. Thermoformed plastic parts can be designed to snap together with few or no fasteners in some cases, allowing them to be compact and sleek looking while still performing the desired task. Plastic is also easy to cut and drill. Some allowed plastics are transparent, allowing you to see "through" your robot to align with dispensers or scoring areas, which can be crucial depending on the nature of the season's game.

Disadvantages of Plastic

One disadvantage of using these techniques is that if you custom thermoform plastic parts, and those parts break during a tournament, you probably will not be allowed to pull out a plastic bender and fabricate a replacement part. Although you may be able to patch up the broken part by using cable ties or bits of metal and screws, we would suggest pre-fabricating a replacement part for any critical item, and bringing that replacement part with you to the tournament.

This requires some judgment. If a part is noncritical and could be easily patched, such as a hopper whose dimensions are not critical, there probably is no need to fabricate a spare. If, however, the part is crucial for some major operation and probably would not work if patched up (such as a custom part used in a shooting mechanism whose dimensions were critical), then you might consider bringing a pre-made replacement part.

Types of Plastic and Their Properties

What types of plastic should you use for different purposes? This section will give you some general guidelines.

In recent years, the types of plastic allowed by the game have widened considerably. The current rules allow for any type of polymer plastic.

Here are major types of plastic that we have found useful:

- ABS (acrylonitrile butadiene styrene) is typically opaque, often black or beige colored. It is easy to cut and drill and thermoforms nicely. It is somewhat soft and can be bent to some degree without heat and still not crack, although in that case it tends to spring back toward its original shape. This property can be used to form spring-like structures in some cases. ABS is one of the materials commonly used in 3D printing. However, we have found that PLA is often a better choice for 3D printing because ABS tends to warp if the part is large. For small parts, ABS is a little more rugged than PLA for 3D printing. Thermoforming temperature is around 325-350 degrees F. Cost is very low, about $2 to $3 per square foot for 1/8" thick sheets as of this writing.
- PETG (polyethylene terephthalate G, also branded Vivex and others) is typically transparent. It is easy to cut, drill, and thermoform and is very good for vacuum forming. In thinner sheets it can be successfully cold-bent. It is used extensively in consumer packaging and plastic bottles and many grades are food safe. PETG has a thermoforming temperature of 250 to 320 degrees F. It is low cost, approximately $3 to $4 per square foot for 1/8" thick sheets as of this writing.
- Polycarbonate (trade name: Lexan and others) is very transparent and extremely impact resistant. So-called "bullet proof glass" is often actually polycarbonate. We do not recommend trying to cold bend polycarbonate as it can shatter (at least one manufacturer recommends a minimum bend radius of 100 times the thickness of the sheet). It is easy to cut and drill and thermoforms well at around 375 degrees F. It is more expensive than ABS or PETG but still moderate in cost, around $5 to $8 per square foot at 1/8" thick as of this writing.
- HDPE (high density polyethylene) is a very rugged, impact resistant, and low friction type of plastic. It is opaque or translucent. Often it is a milky white color. Because it is used for plastic cutting boards, you can sometimes find it in many different vibrant colors as well. HDPE is easy to ma-

chine, cut and drill. It thermoforms well. It is relatively low cost. We use HDPE quite a lot on our team. It is wonderful for creating custom brackets, gripper jaws, and other purposes. If the part you need doesn't require transparency, it is a great choice.

- LDPE (low density polyethylene) is very similar to HDPE but is a little more flexible. So for a bracket that needs to stay stiff it is not as good a choice. It is about the same price as HDPE and other characteristics are very similar to HDPE. It does not come out as "clean" when machined as HDPE does, so we prefer HDPE in most cases.
- Polypropylene is extremely similar to LDPE and effectively can be used for the same purposes.
- Acrylic (polymethyl methacrylate, also known as "Plexiglass") is transparent and available in many colors because it is extensively used in the sign industry. It is very tempting for your team to use acrylic because it is commonly available at hardware stores. Please do not use it for any structural part. It cracks and breaks very easily compared to alternatives such as PETG and polycarbonate. It may be used for decorations and very low stress components, but there are better choices.
- PLA (polylactic acid) is commonly used in 3D printers. From our experiments, it is better for most purposes than ABS when used for 3D printing. This is because it has very little tendency to warp on the print bed. This allows you to create large parts without a heated print bed. PLA is made from corn and is biodegradable, so it's more "green" than other plastics made from petroleum products. Other than 3D printing, PLA is not used much by our team because there are lower cost alternatives. PLA is a little bit on the brittle side so you must properly reinforce it with other plastics or metals if you have made a large 3D printed part that will undergo stress.
- KYDEX (Acrylic/PVC blend). Kydex is a brand name for an Acrylic/PVC blend which is typically opaque (often white, grey, or black). It is specifically formulated for good machining and thermoforming characteristics. It is rigid and we don't recommend cold bending, but the rigidity is an advantage for many uses. Thermoforming temperature is 325 to 390 degrees F. Some formulations are also chemical and flame retardant. It is used extensively in commercial products such as printer housings. But all these properties come at a cost: it is among the most expensive plastics allowed by recent FTC rules at $9 to $11 per square foot for a 1/8" thick sheet. Because of the cost, we would recommend some of the other plastics unless

you can take unique advantage of what Kydex has to offer or unless you would exceed maximum allowed sheet sizes of the other plastics.

- PVC (polyvinyl chloride). In recent years PVC pipe has been allowed in various sizes, but not PVC sheet. Note that PVC pipe fittings (such as end caps, t-connectors, etc.) have not been allowed, only PVC pipe. There have been length and diameter restrictions as well. PVC was extensively used in the 2009/10 season for various "shooter" mechanisms because the game involved shooting wiffle balls which happened to fit very nicely in 3" diameter PVC pipe. PVC is easy to cut and drill and PVC glue is commonly available and extremely strong. Because only PVC pipe has been allowed recently and not sheet, we would not recommend thermoforming, although PVC pipe can be thermoformed to some extent.

The rules change every year and new kinds of plastic may be allowed (or some of these may be disallowed) in future games. The quantity and maximum thickness allowed may also change from year to year. As of the 2013/2014 season any thickness of practically any common plastic is allowed. In past years some very common plastics such as acrylic have not been allowed, so be sure to check the rules and make sure you are buying the correct type.

Where To Purchase Plastic, Plastic Accessories, and Plastic Forming Tools

 # 132. TIP: US Plastic corporation (www.usplastic.com) is a good place to purchase plastic sheet.

They have a wide selection of PETG, Polycarbonate, ABS, and other plastics. Also they sell plastic glues and tools.

 # 133. TIP: McMaster-Carr is a good place to purchase plastic sheet.

They have a good selection of PETG, Kydex, ABS, and Polycarbonate plastics and glues.

 # 134. TIP: Amazon.com (www.amazon.com) is a good place to buy 3D printer plastics (ABS and PLA).

Free shipping using Prime and low prices from a variety of vendors have made Amazon.com our supplier of choice for 3D printer filament plastics.

 # 135. TIP: Plastic benders of several sizes can be purchased from Delivies Plastics (www.deliviesplastics.com).

They have several sizes of "strip heaters" to meet any budget. You can make very useful plastic parts even with their 12" size for under $90 as of this writing, but we would recommend the 21" size if your budget allows ($120 as of this writing). Most parts you will want to fabricate will be less than 21" long so this is a good size.

They also sell plastic welders, but note that welding plastic has *not* been allowed by the rules in recent years, so we do not recommend these unless the rules change.

 # 136. TIP: Plastic benders and heating elements can be purchased from TAP Plastics (www.tapplastics.com).

More expensive than the Delvies strip heaters, but very high quality, is the TAP Plastic 24" Free-Standing Heater for about $240 as of this writing. However, they also sell a 36" heating element for about $70 and they offer video instructions on how to turn this into a low cost strip heater using materials like plywood, aluminum foil, and fiberglass tape. An adult should closely supervise any project to build a strip heater!

Plastic Fabrication Techniques

This section will provide tips on various plastic fabrication techniques. By following them, you can produce professional-looking plastic parts safely.

Fabrication techniques for plastic include common shop operations like cutting and drilling, but we will focus in this section on operations like thermoforming and gluing that are more unique to plastic.

Thermoforming is the process of heating plastic to its thermoforming temperature (250 to 400 degrees typically for the types of plastic used in FTC) and then bending the plastic into a desired shape. Thermoforming is typically performed using either a heat gun or a strip-heater.

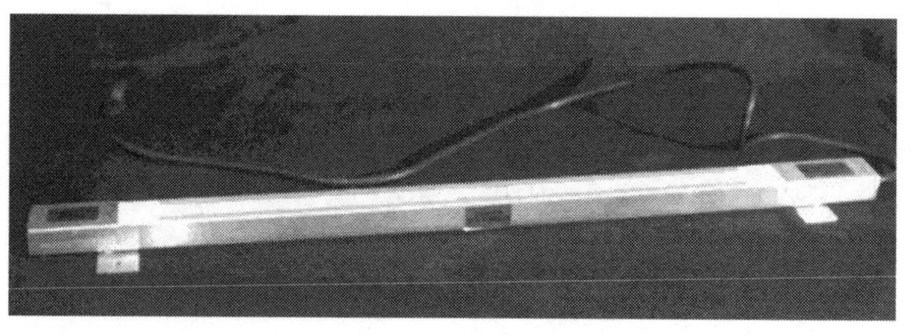

Heat guns are low cost and can be used to make many kinds of useful shapes. But they are not precise and often result in "wavy" looking plastic because they heat large areas. A much better way to make straight bends in plastic is to use a strip-heater, shown in the picture above. A strip heater allows you to carefully heat up just a narrow strip of plastic, allowing for a nice clean bend. By planning out your bends, you can make professional looking plastic parts using a strip heater.

Some plastics can also be formed using cold-bending, especially if the sheet thickness is under 1/16 inch. However, you have to be very careful when cold-bending to avoid shattering the plastic. As noted in the discussion of plastic types above, some plastics are better for cold bending than others. Cold-bending can be performed using clamps, or even just pliers for small parts. Be sure you are, as always, wearing safety glasses in case the plastic breaks while bending. Also, allow for the plastic to somewhat "snap back" after the bending operation. You may have to overbend the piece to take into account the snap back action.

General Plastic Fabrication Tips

 # 137. TIP: Many plastic fabrication techniques are illustrated in free instructional videos at www.tapplastics.com/info.

They have videos on cutting, bending, polishing, and more! Note that most of their information is general and applies to any thermoplastic, but some are specifically about acrylic plastics which are illegal in recent FTC seasons.

 # 138. TIP: When using cement with plastic, be very careful to obtain a cement compatible with the type of plastic you are using.

Recent FTC games have allowed cementing plastic parts as an acceptable fastening method.

When cementing plastics, you must make sure you are using a cement compatible with the specific type of plastic you are using. The best types of cement to use are "solvent" glues. They work by actually chemically dissolving the plastic and allowing it to weld to itself. For example, Weld-On #3 is a solvent cement that works with polycarbonate and PETG. Weld-On ABS 771 works on ABS. There are many other brands, these are just examples. Search some of the plastic supplier sites mentioned earlier to find compatible cements. View the instructional videos at TAP Plastics to see how to properly apply solvent cements.

 # 139. TIP: When using heat to thermoform plastics, make sure you set up in a safe place, away from people and flammable objects. Warn others around you that heat is about to be used. Adult supervision is required.

As with any tool, improper use can cause accidents. Strip heaters and heat guns are—as their name implies—hot! This means you must set up your project in a safe place, inform all team members about the possible danger, and make sure an adult is fully aware of what you are doing and is actively supervising.

 # 140. TIP: Before using heat in the workshop, make sure fire alarms are operational and have a fire extinguisher nearby.

The time to make sure safety equipment is working is before there is a problem, not after!

 # 141. TIP: Be certain a strip-heater is turned off immediately after use. Use an appliance timer as a fail safe way to ensure a strip-heater is not left on.

Strip heaters can reach 900 degrees F. or more and you must be certain that they are unplugged after use. A fail-safe way to ensure you do not accidentally leave a strip-heater turned on after a build meeting ends is to always use an appliance timer with an automatic OFF feature. An example of such a timer is the Simple Touch C30002 which automatically turns off 60 minutes after its "on" button is pressed. It costs only about $15 as of this writing. There are other similar products with different features, so do your own research. Do not use timers that come on at certain times of day, use ones that turn on for a specified period then turn off and stay off until manually reactivated! (We also recommend the use of safety timers like this for soldering irons and battery chargers.)

 # 142. TIP: When using heat (heat gun or strip-heater) to thermoform plastic, wear heat resistant gloves.

When bending larger parts it is easy to hold the plastic in an area away from the heat, but if your hands are going to get within a couple of inches of the heated part, you should wear gloves. If the part is very small, you should use pliers to hold it. Also remember that the part remains hot for some time after the bend is complete.

 # 143. TIP: Plastic glues require proper ventilation.

Please read carefully manufacturer instructions when using plastic glues. Most of them require good ventilation. Never keep glues or other flammable materials anywhere near places where you do heating operations like soldering or thermoforming!

Bending Plastic Using a Heat Gun

 # 144. TIP: A heat gun can be used to make minor bends on relatively large parts, but the results often look "wavy."

Heat guns offer a very low cost way to make some types of plastic bend. Generally speaking, the results will not look anywhere near as professional as using a strip-heater, however, and the bends will not be as precise. Sometimes a heat gun can be used to adjust a part that was formed using strip-heating.

For example, let's say you make a bend in a part using a strip heater, then once the part is on the robot you find it should have been bent a little more acutely. Instead of removing the part from the robot and taking it back to the strip-heater, you may be able to hit it with a heat gun for a few seconds then slightly adjust the bend right on the robot (this is desirable because you won't need to guess how much farther to bend it, you can bend it just until it fits where you need it to be). Be sure, of course, that you do not expose electrical parts of the robot to excessive heat in the process. Don't ever do this near the battery, Samantha module, or NXT Brick, for example.

 # 145. TIP: Make a frame to support the plastic you wish to bend using a heat gun, use that frame to make the shape as precise as possible.

If you want to shape a relatively large piece of plastic into a complex shape such as a "hopper" or a "slide" to deliver game pieces, you may get a better result by first creating a frame-like structure that is the same shape as what you are trying to fabricate. For example, use plywood or even cardboard and tape to create a kind of mold for the shape that you are trying to produce. Place the plastic over this frame and heat up small areas at a time using the heat gun. The plastic will drape down over the frame in approximately the desired shape. Again, this generally does not work as well as a strip heater and usually results in some "wavy" characteristics in the final product, but it should come out better than just heating and bending "by eye."

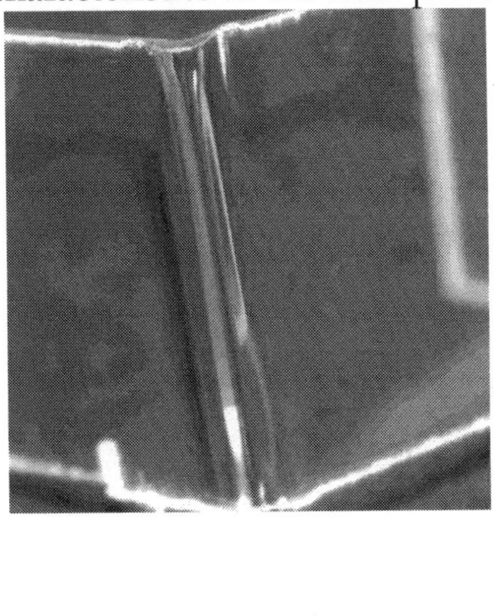

Bending Plastic Using a Strip-Heater

 # 146. TIP: Allow sufficient time for strip-heaters to heat up.

Strip-heaters take a long time (around ten to twenty minutes) to heat up, so plan ahead if you know you will need them. If you try to use a strip-heater before it is hot enough, your bends will take a long time to make or will not turn out well.

 # 147. TIP: Use pliers when bending a small part.

You can often get a better, sharper bend on small parts if you use pliers instead of your hands. If you are concerned about marking the plastic with the pliers, you can purchase nylon tipped pliers. For large parts using a strip-heater things usually work fine using just your (gloved) hands, because only the section near the strip reaches its thermoforming temperature. But small parts will often heat up such that much of their area becomes soft, so the pliers allow for a tighter bend.

 # 148. TIP: The plastic is ready to bend when it just starts to become limp. Practice on scraps first.

After placing the plastic on the bender, keep one hand on each side of the heating strip. Every now and then test whether the plastic is ready to bend by putting a little pressure on it (bending away from the heating strip). When it starts to feel like it is giving way, test the plastic more frequently, such as every few seconds. You will feel the plastic suddenly go "limp" and bend very easily. At that moment, move it away from the heat and bend to the desired angle. Hold it firmly and allow it to cool and become stiff again.

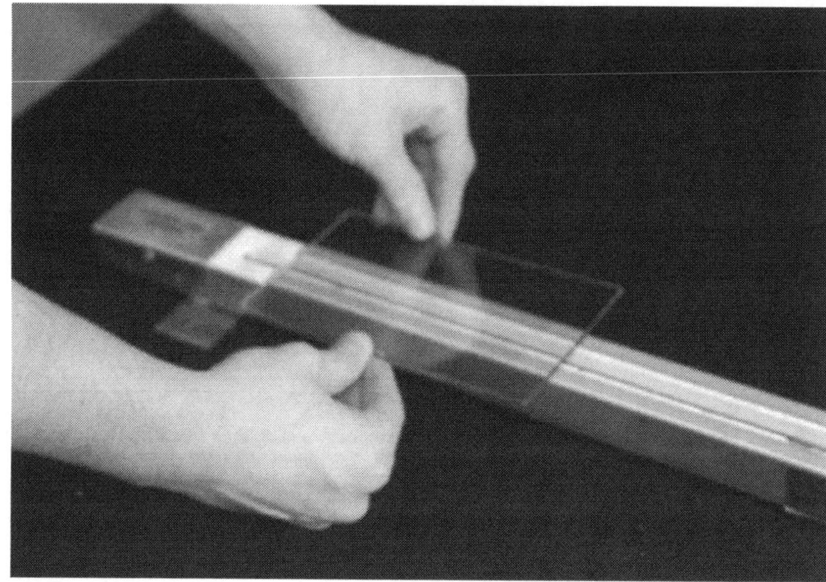

 # 149. TIP: If bubbles form in the plastic, you have overheated it. Heat it less next time.

The picture here shows plastic that has been overheated. Bubbles form in the bend. The part may still work just fine, it just won't look as nice. But next time, learn to bend it earlier, before this occurs.

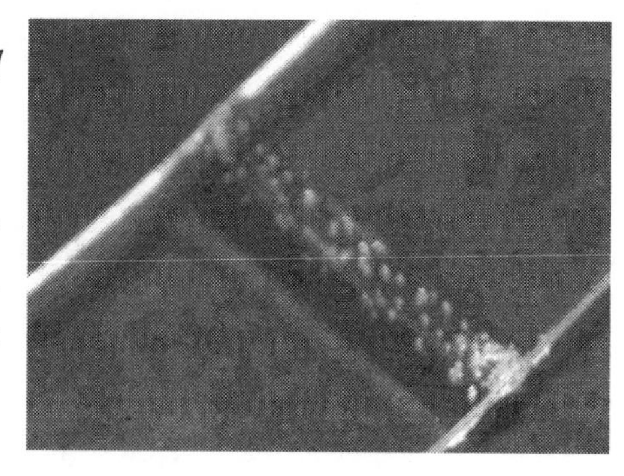

 # 150. TIP: If "crack" marks appear in the plastic after a bend, it was not hot enough, Let it get a little more limp before bending next time.

If you see crack marks (also called "craze" marks) in the bend after cooling, this means you forced the bend before the plastic was fully heated. This causes stress in the plastic and little cracks appear after cooling. Next time let the plastic heat up a little longer. If you find yourself putting a lot of force on the plastic to bend it, it's not hot enough. It should be quite easy to bend once the thermoforming temperature is reached.

 # 151. TIP: Plan out your bends using posterboard or paper prototypes.

It is possible to make very cool looking shapes and even snap-action parts by bending plastic. But don't waste expensive sheet plastic by trying to "wing it." Instead, prototype your new plastic part by cutting out poster board or paper versions. Fold your prototype the same way you plan to "fold" the plastic using bending operations, and then make sure the result fits where you want it to fit on the robot. Use this paper version as a template to mark where the plastic needs to be cut and bent. We use dashed lines to mark bends and solid lines to mark cuts.

3D Printing Plastic Parts

3D printers have become far more affordable in the past few years and many teams are now using them to fabricate some parts. While an entire book could be devoted to just this topic alone, here are a few tips our team have found to be useful.

 # 152. TIP: Design parts to be thick enough to withstand stress, and take into account the "grain" of the printed piece.

3D printed parts can be easier to break than plastic parts made from other fabrication techniques. This is because the 3D printed part is built up with a series of layers. Those layer boundaries can be a weak point, and 3D parts may break more easily along them, just as it is easier to split a log by slamming your ax along the grain rather than across the grain. By rotating the part to be printed (if the result is still printable of course) you may be able to position the "grain" of these printed layers in such a way that expected stress forces are "against the grain" allowing for a stronger piece.

 # 153. TIP: Reinforce 3D printed parts with thin layers of PETG.

Another way to handle the "grain" issue mentioned in the prior tip is to reinforce your 3D printed part using a very thin layer of PETG plastic. If you purchase a sheet of PETG that is in the range of 0.02 to 0.04 inches thick, it can easily be cut with ordinary scissors or a razor blade. You can cut out an outline of transparent PETG that matches the shape of one face of your 3D part. Then attach it using glue or even fasteners like screws. This thin layer of PETG will accomplish 2 things. First, it will tend to stop the 3D part from cracking along a layer boundary. Second, it is nice and shiny and actually can make the part look more impressive.

 # 154. TIP: A great place to get 3D printing information, software, and models is Thingiverse (www.thingiverse.com).

Thingiverse is a community for 3D printing enthusiasts. You will find thousands of models for parts that may be usable in FTC. Often the source files are also posted so you can modify parts for your purpose. In addition, there are some kinds of file that allow you to "customize" parts. For example, you will find a gear generator that could make gears that are quite useable in low-stress situations. You can even post your own designs to help out other people trying to solve problems using 3D printers.

 # 155. TIP: Design parts whenever possible to print with few or no supports.

When you design a CAD model for a 3D part, it is important to bear in mind how the printer works. A part that has overhangs with nothing underneath will need "supports" in order to print. We have found that you are far better off designing the part so it has at least one orientation that allows it to print without supports. Supports add a lot of time to the printing process. When a support is needed, it is often superior to build the support into the 3D model rather than using automatic support generation tools included with most 3D printers. This is because you can design a minimal support that is a "break away" part that cleanly separates from the model.

 # 156. TIP: Instead of making one large part as a single unit, divide up a large part into several pieces and place screw holes in the right spots for fastening.

Large parts can take a very long time to print. We have had parts that require 15 hours or more. But the longer a print takes, the more chance something will go wrong. Snagged filament, alignment issues, or a thunderstorm knocks out your power! It is often easier to divide up a part into several subassemblies and print each piece in a smaller job lasting just a few hours. This also has the advantage of allowing for support-free printing by carefully choosing how the larger piece is broken up. In addition, damage can be repaired in some cases by just replacing a single piece of the assembly rather than doing another massively long print job!

 # 157. TIP: For parts with critical dimensions, print a spare and bring it with you to the tournament.

If a 3D printed part does not have critical dimensions you can repair it at a tournament using tape or by screwing additional pieces of plastic into it. But if it's dimensions do need to be precise then repairs may render it useless. Since you will not have time at a tournament to print a new part (even if you had a printer in your pit) you should seriously consider printing a spare.

Materials and Compliance

 # 158. TIP: Be extremely careful to understand the materials limits, types of materials, and allowed dimensions.

The rules are very strict regarding both the types of materials and their dimensions (length, width, and thickness). It's very important to read and understand the rules to avoid violating these limits, which could result in your robot failing hardware inspection. If you are running low on one material you can redesign to use another that you have more of. But be warned: the rules change from year to year; make sure you understand the current rules.

 # 159. TIP: Keep a folder or binder that contains all receipts for materials you used so you can prove you purchased the correct thickness and type of material.

During inspection at a competition you will be asked questions on whether your robot uses the correct materials, it is a good idea to have proof you purchased the correct materials.

 # 160. TIP: Create a "cut sheet" for each part showing how it was cut from the allowed material.

In prior seasons this was necessary to be sure that the amount of material used is within the specified limitations. In recent seasons most material limits have been removed and currently cut sheets are no longer required. However, it may still be a good idea for you to keep cardboard templates for all your cut parts, just so you can fabricate new versions in case of damage. The simplest way to create a cut sheet is to obtain a 24" square of poster board or large paper, one for each material (plastic, aluminum sheet, etc.) and just trace each part made onto the sheet for the corresponding material.

Evaluating and Debugging Hardware

 # 161. TIP: Build a full field if possible, early in the season. If you don't have room, at least build a half-field.

Championship teams almost always have a field of their own to practice on. If you don't have the space for a full field, a half field is still a big advantage. If you don't have the resources for even a half field, try to find another team in your area and offer to share expenses or labor to build a joint field.

Without at least a half field, your drivers cannot really get a feel for how to play the game, and programming a competitive autonomous mode is nearly impossible.

 # 162. TRICK: Buy the official game elements, do not substitute!

Frequently FIRST will designate a specific vendor and model number for the game scoring elements. It is very important that you buy the official elements and do not substitute locally available items that seem the same.

In the 2009/10 season the scoring element was a wiffle ball. Early in the season we bought some random wiffle balls just to test early prototypes. A few weeks into the season FIRST announced a vendor and model number for official game elements. We purchased a complete set of the official wiffle balls and found they behaved very differently than the early ones we purchased. The exact configuration of the holes in the balls, the precise diameter, and the weight: all of those things, although only different by tiny amounts, caused a huge difference in performance of the robot.

 # 163. TRICK: Design repeatable tests for all major subsystems. Document baseline tests in the Engineering notebook. Compare improvements with the baseline and document them as well.

For example, in the 2009/10 season we designed a shooter to propel wiffle balls through a goal. Our baseline test was to set the robot against one of the field walls then repeatedly fire balls, marking where each one landed with a bit of tape. The most desirable result would have been for every ball to land in exactly the same spot, of course, but in reality the landing points were scattered. One test of shooter accuracy was to simply measure the diameter of the smallest circle that could enclose this scattered set of points. The smaller the scatter diameter, the better.

The first photo shows the overall setup. In the second photo, a set of tape marks from a test has a diameter about the same size as the 18" ruler shown. As we tried ways of improving the accuracy, we could repeat the test and see if our changes really worked or not.

Another test was to fire as many balls as possible in a set amount of time. Firing rate was affected by hopper jams and other issues. As we made improvements, we could re-run these tests to see how effective they were.

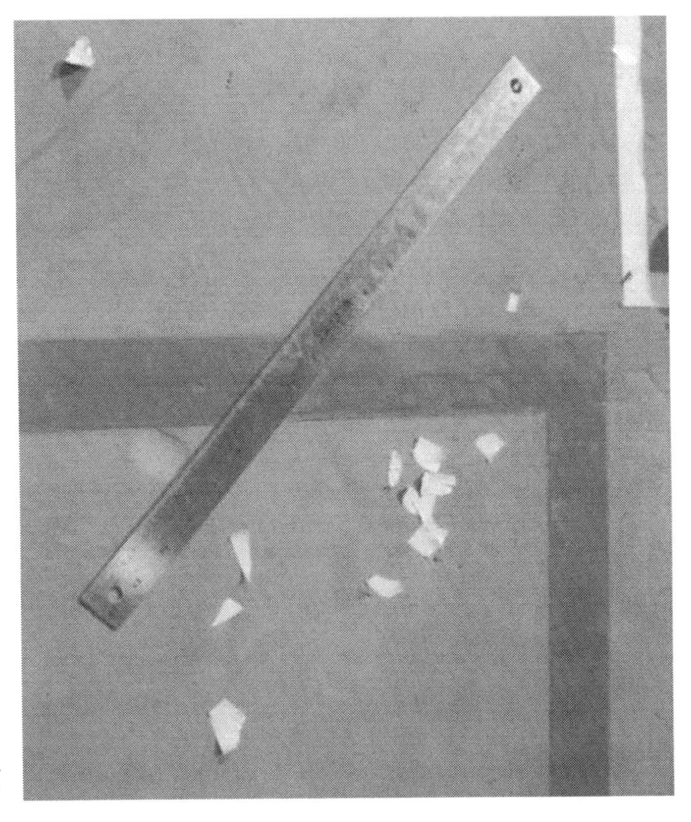

In one case a completely new design was made for the piston (the part that struck the ball) on the theory that some inaccuracies in the piston were causing misfires. It turned out that testing proved the old piston was in fact clearly superior to the new piston, so we went back to the old one. Don't just assume that rebuilding something will result in an improvement; sometimes it does not for reasons that are so subtle you may never completely figure them out. Testing is the only way to prove whether one design or another is superior.

Documenting all these tests in your engineering notebook also impresses the judges: you are using real techniques used in industry to find problems, measure results, and make improvements systematically. You're not just "winging it."

 # 164. SECRET: Think in terms of failure modes: all the ways a system could fail. Observe repeated trials to find all the failure modes; don't assume there is only one. Analyze to determine which failure modes are the most prevalent; concentrate on fixing them first.

Let's say a certain mechanism works sometimes but not others. The question is: why doesn't it work all the time? Sometimes there may be a single cause for failures. In other cases, there may be several different causes. Each of the causes of failure is called a *failure mode.*

For example, say you have a harvester that picks balls off the field floor, as was common in the 2009/10 season. Perhaps it usually works well, but sometimes a ball fails to be transported all the way to the hopper. Careful observation might reveal that there are two different reasons why the ball may fail to be transported. Perhaps one reason is that one side of the harvester is too narrow to accommodate the diameter of the ball causing it to get stuck, and perhaps another is that there is not enough traction in a certain section of the conveyor mechanism to lift the balls. Further, you conduct tests and find that the second failure mode occurs far more frequently than the first, and is far simpler to fix by simply adding some nonslip pad for extra traction. The other failure mode will require the entire chassis to be disassembled to widen the constricted area.

By doing such analysis you can devote your build team resources to fixing the most frequent failure modes first. If, in the example above, you had a scrimmage the very next day, there may not be time to disassemble the entire chassis to fix the first failure mode, but because you did some analysis you determined that the build team can fix the most prevalent issue with time to spare. The team can schedule a more extensive overhaul of the chassis after the scrimmage to correct the remaining failure mode. You warn the drive team that harvesting on one side of the harvester may occasionally cause stuck balls and ask them to try to compensate by steering balls in to the other side whenever possible. This maximizes your chances of success given the constraints on time and resources.

 # 165. SECRET: Don't "solve" problems that don't exist. Focus on the real bottlenecks.

It may seem obvious that you shouldn't solve a non-existent problem, but we see it all the time. One recent example involved a student who wanted to build an elaborate system to make our wiffle ball shooter fire faster and to vastly increase the size of the hopper. The coach asked him if we ever had a problem firing every ball in our hopper before time ran out, to which the student had to answer, "no." Then the coach asked if our hopper was frequently full to capacity. Again, the answer was "no."

In fact, the bottleneck was not the firing speed, and the bottleneck was not the hopper capacity; the bottleneck was the speed of the harvester. Of course it would be nice if the gun fired faster, and it would be nice if the hopper could hold more balls, but neither of those "fixes" would have improved the number of points scored, because the harvester could not even keep the existing hopper filled fast enough and time rarely ran out with balls left unshot.

The point is, there may be lots of things that are nice to improve, but identifying and fixing the *real* bottlenecks in the system will actually make you score more points and win more rounds of play. Once you fix one bottleneck, the bottleneck then may move to some other part of the system and require further analysis. But if you fix the bottlenecks in the order that maximizes your scoring capability then you'll have more success.

 # 166. SECRET: Do slow motion camera studies to determine what failure modes exist; use the data to design improvements.

Sometimes a mechanism acts so quickly that it is difficult to observe why it is occasionally failing.

Using any standard camcorder you can record fast moving mechanisms like shooters and then look at the results frame by frame in video editing software to find problems. We frequently do camera studies to determine what is causing repeatability problems. Standard cameras give you 30 frames per second of detail, or about 33 milliseconds per time slice. Many HD camcorders record at 60 frames per second for even more detail. There are even some high end camcorders

now, still within many people's budgets, such as some Sony models, that record at 240 frames per second(smooth slow motion mode) for a 3 second burst, about 4 milliseconds per frame!

In the 2009/10 season our wiffle ball shooter had a 0.2 second cycle time. From the time the piston was released to the time the ball exited the barrel was only 80 milliseconds. We could not tell why it was sometimes misfiring, causing about one ball in three to travel only half the usual distance. Fixing this problem would improve our scoring by 33%, a huge margin!

We made dramatic improvements to our gun accuracy by slowing down the action and figuring out what was different on "good" shots versus "misfires." In this case the problem was trivial to fix: the holes of the wiffle ball game elements were sometimes getting snagged on a stopper that prevented the ball from rolling back against the piston. Changing the shape of the stopper solved the problem.

Miscellaneous Building Tips

167. SECRET: *Your robot is never "done" until the final game of the final post-season match. Keep improving it all season.*

The best teams don't rest; they keep developing their robots throughout the season. You may only have the robot barely functioning and you may have no autonomous mode by the time of your first scrimmage, for example. You should take whatever you learn from that first scrimmage and design some improvements, or in some cases entirely new mechanisms. If your region uses qualifier tournaments, you should try to have a robust autonomous mode by then so you can try it out against real competitors. Again, you will learn more about the game, and your team must continue to brainstorm new improvements before your regional Championship. If you are lucky enough to win your Championship or otherwise qualify for the World Championships, keep pushing your design, because you'll need to bring your very best work to compete against the best in the world!

We even continue to make minor adjustments after the World Championships for post-season tournaments. Even though such tournaments are not official, you can learn things that may help you compete in the next season's game.

168. SECRET: *To clamp something between two hubs, drill out the holes on one hub, insert the screw through that one first, and then use the threads in the other hub to tighten.*

There are times when you want to clamp something, perhaps a metal bar for use in an actuator, between two hubs in order to get better stability and allow two set screws to share the load (avoiding a single point of failure as well).

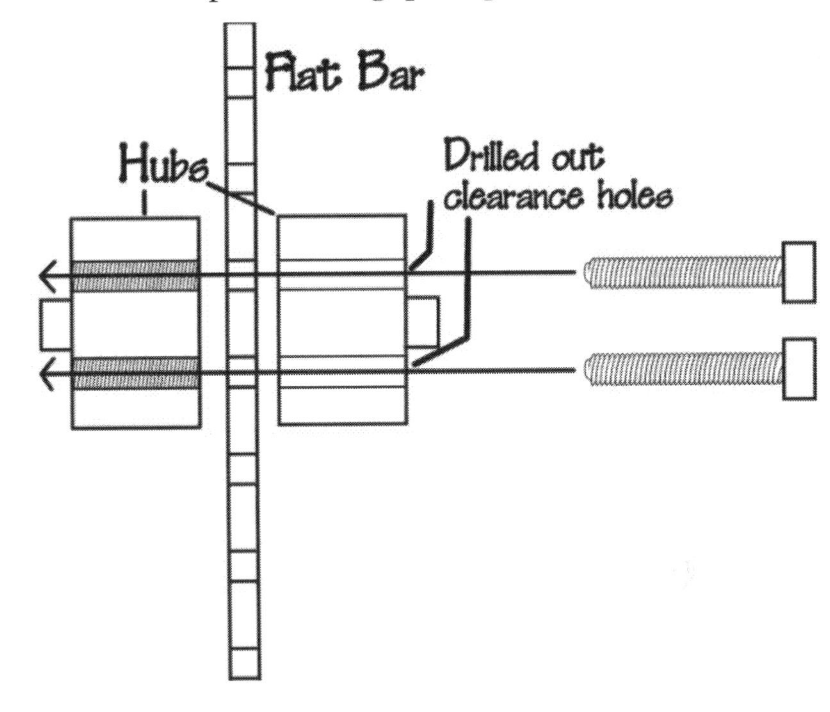

If you try to do this, what you'll find is that you can't get both of the hubs tight against the object you're trying to sandwich between them. This is because the hub holes are threaded and the screw will go through one, then through the item you are clamping, but won't align perfectly with the threads of the hub on the other side.

The simplest solution to this is to drill out the holes in one of the hubs. Use a drill bit just barely larger than the screw diameter (0.14"). The screw first goes loosely through the drilled out hole, then through the item you are clamping, then threads into the second hub on the other side.

 # 169. TIP: Motor shaft hubscan make good spacers for axles.

Recall that one of our prior tips warned you not to use the wrong kind of hub. Motor shaft hubs are for motor shafts and axle hubs are for axles. But sometimes you need a spacer that is much larger than the white nylon spacers normally used on axles, and you may want the spacer to be able to spin freely on the axle. In this case, a motor shaft hub can sometimes make a perfect spacer, without using the set screw to allow it to simply spin on the axle. Its larger center hole allows free motion.

 # 170. TIP: Tubes are notoriously fragile; use tube plugs and a strip of aluminum to ensure tube clamps don't crush the end of the tube.

The Tetrix tubes are not a good choice to bear significant weight, but they are often essential to form "rollers" for harvester mechanisms. One frequent problem is that the tube clamps crush the ends, after which they work themselves loose. In the 2009/10 season Tetrix introduced "tube plugs" to help reduce this problem.

A tube plug fits on the end of the tube, then the tube clamp cannot so easily collapse the end of the tube. However, another problem arises: the tube plug itself can slide down the tube, again allowing the end to collapse. One technique to deal with

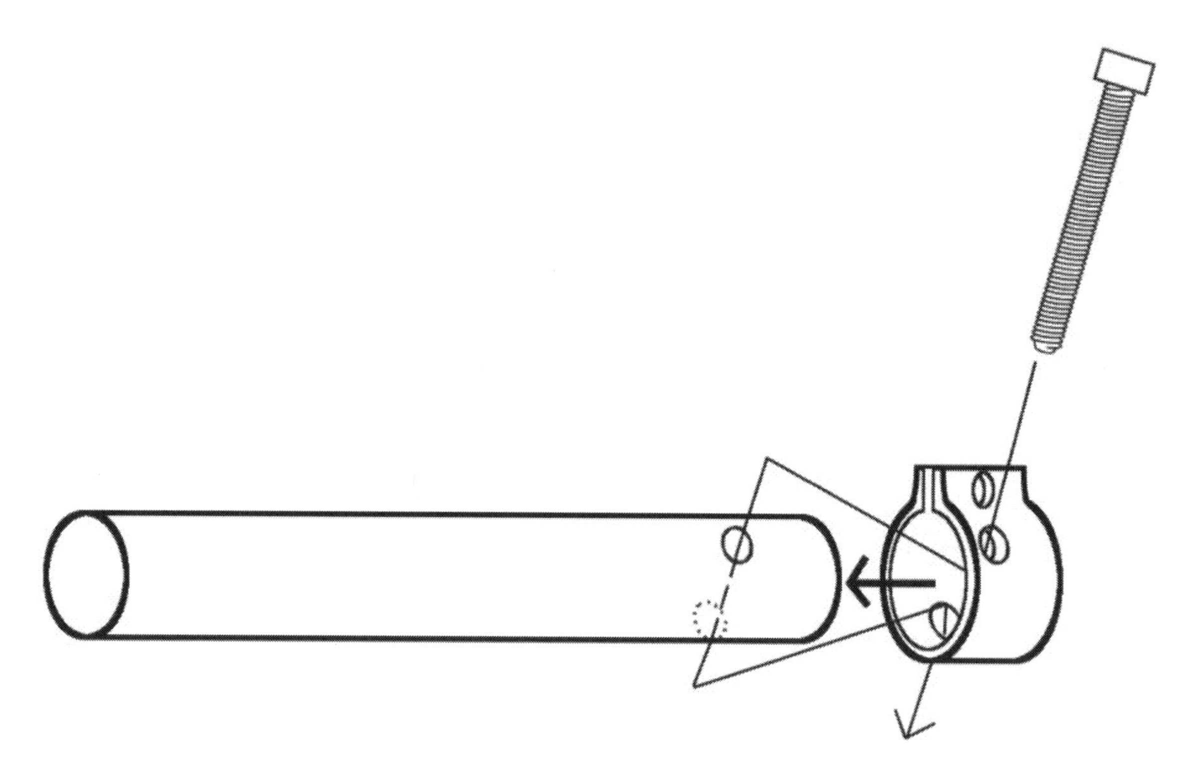

this problem is to insert a strip of aluminum inside the tube, making the strip just long enough to hold the tube plugs on each end of the tube in place. You may need to cut the strip of aluminum slightly longer than necessary then file it down, testing every now and then by inserting tube plugs on each side, until the plugs are both flush with the end of the tube.

171. TIP: *Another technique to secure a tube is to drill a hole through the tube and tube clamp then use a long screw to secure it.*

When you need even more holding power than the prior tip provides, drilling a hole clear through the tube clamp, tube, and tube plug can work. The only tricky part is that there is not a lot of room to drill the hole in the clamp and there is very little metal on the end of the tube to make such a hole. For this reason, it is still not a good idea for such a tube structure to bear more than minimal lateral forces.

 # 172. TIP: If you secure all four screws to a split clamp, you will not be able to fully tighten the clamp on the axle. Instead, leave one of the two screws near the split part out.

The split clamps provided with the Tetrix set can be frustrating. What you will find is, if you first affix the clamp to a wheel or gear using all four screws, you can no longer tighten the "clamp" part onto the axle, because the four screws actually hold the split part open regardless of how tight you try to make it. You can alleviate this problem by leaving one of the screws next to the "split" off. Three screws is normally quite sufficient to hold the wheel or gear in place.

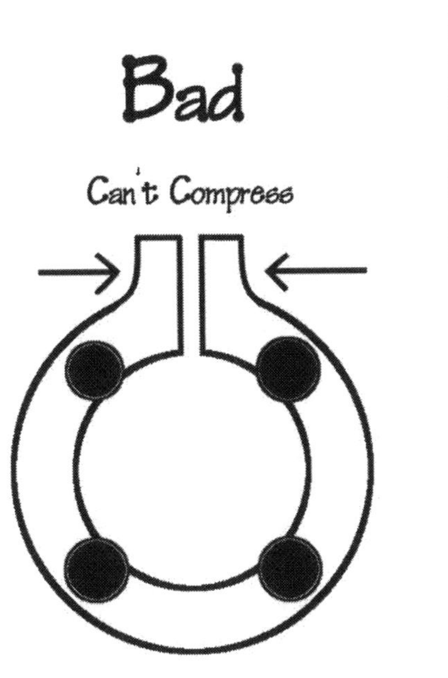

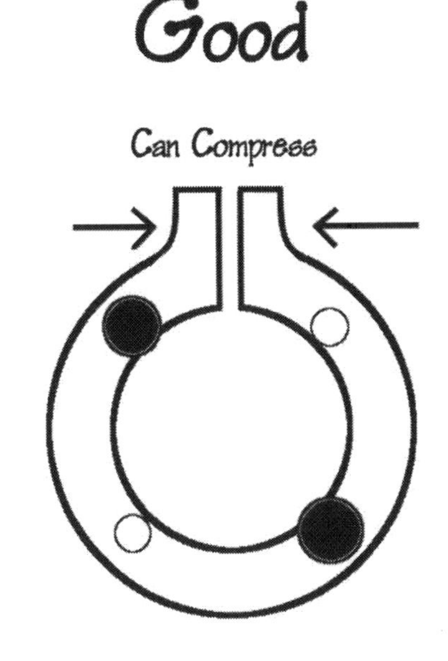

Another technique is to tighten the split clamp onto the axle before affixing it to the wheel or gear.

Part IV: The Samantha WiFi Module

The Samantha WiFi Module was introduced in the 2010/11 season and replaced Bluetooth as the communications method used by the Field Management System (FMS), although in prior years it was called the Field Control System (or FCS) so many people use the terms interchangeably.

Being a new system, there were inevitable problems that had to be worked out over the course of the season. Several of these problems could be reduced or eliminated by proper mounting of the Samantha module on the robot, others required site specific preparation activities, and still others are really not entirely clear as of this writing.

This section outlines best practices for the use of the Samantha module. Although following these tips should reduce WiFi issues and represent your best chance of keeping your communications alive, it is still unclear what level of problems will exist in any future seasons. FIRST is working very hard to identify all of the failure modes and is taking actions to correct problems. We would urge you to carefully read all communications from FIRST on this topic that are posted during the coming season and comply with any new recommendations they release.

 # 173. TIP: Mount the Samantha Module high up on your robot with the lights facing out to the side and the wires coming out the bottom toward the floor.

The antenna is inside the module between the NXT light label and the FTC logo. It is best for it to be oriented as shown in the picture: with the power and USB cables coming out the bottom toward the ground, and the lights facing out the side of the robot. Mounting the module this way also makes it easy to see the lights.

 # 174. TIP: Do not mount the Samantha module near too many metal parts.

While it is ok to mount the module to a channel for stability, do not bury it among large masses of metal. We have seen modules mounted inside the slot of a channel, which is actually a very bad idea!

 # 175. TIP: Use a short, six inch, USB A-A extension cord attached to the Samantha module, and use cable-ties to secure the USB cord against movement.

The reason for this is that the first version of the Samantha module had a manufacturing flaw whereby the USB connector was not sufficiently mechanically connected to the circuit board. This causes the module to lose the USB connection if the USB cable moves in relation to the module's USB socket. For this reason, it is recommended that a cable tie secure the USB cord near the module, using a hub as a spacer. But because this makes it difficult to plug anything into the Samantha module (for example a flash drive to program the module's security settings for the FMS), the short USB extension cable is used. That way, you merely unplug the other end of the extension cable to flash the module.

In the photo, note the cable tie securing the short USB extension cable around a hub (which is being used as a spacer). The other end of the short extension cable is easily accessible, allowing the module to be flashed at a tournament.

Short USB extension cables can be found at electronics stores or even Amazon.com by searching for the term: "short USB A-male to A-female cable," or sometimes also: "USB A-A extension cable." They only cost a few dollars.

 # 176. TIP: Wire the Samantha module in parallel with the first motor controller. You may either solder or use crimp connectors.

Another issue with Samantha in its first season was that if the module ever sees voltage from the battery drop very low, then it cuts off the USB connection (which cannot operate below about 5 volts). The problem is that if the module is wired into a motor controller, a collision may cause the voltage to appear to drop below this threshold momentarily (due to wires shaking around). The effect was that the Samantha module would lose connection for approximately 10 to 15 seconds as it rebooted itself.

Often you would see robots start spinning around in circles when this occurred. As the Samantha module rebooted and re-established connection to the NXT brick, the brick would happily just continue executing the last command it saw come in. If you happened to be turning when communications were lost, well, the robot would just keep on turning. The same would occur if the robot happened to be moving straight ahead; it would just keep going, plowing right into field elements or other robots. (See the later in this section for a tip that will allow your program to detect this situation and turn off drive motors to avoid burning out motors.)

The solution to this problem is to make sure the module has a direct, firm connection to the battery which is totally independent of the motor controller terminals. In other words, you take the red wire coming from the kill switch and you solder or crimp connect that directly to both the Samantha module's red wire as well as another red wire in a "Y" configuration. You run that new red wire to the first motor controller. Same goes for the black wire coming from the battery. Under this configuration, even if the motor controller was removed entirely, the Samantha module will continue to receive full battery power.

 # 177. TIP: Add your team logo to the Samantha FMS's timer screen

You can add your own team logo to the timer screen on the Samantha FCS by following these steps:

1. Create the logo. It must be a 64x64 GIF file, at 72 DPI (animated gifs are not allowed).
2. Connect to your NXT in RobotC.
3. Go to Robot/NXT Brick/File Management. Select "Download." Select your logo file, and click "Open"

4. The next time you connect to your robot in the FMS, click on the yellow “details” button (the yellow color means that a logo was found), and check the "Display Logo On Timer" check box.

The team logo will now be displayed if you open the “Timer Display” window.

Step #4 has to be redone on every computer on which you want the logo to be displayed.

Note that the field tech advisers don't appreciate it if you ask them to approve your logo when they're very busy. Be respectful of their time at busy tournaments.

 # 178. TIP: Download the Samantha Documentation and keep it on your laptop.

FIRST makes all of the documentation for the Samantha Wifi Module available for free, on the same page as the FMS. Download it to your hard drive, it can be useful (and you rarely get internet access at competitions).

 # 179. TIP: Always test over Samantha before a tournament.

Always run your real driver training sessions using Samantha, do not use Bluetooth.

While it is more convenient to test over Bluetooth when you are making a lot of changes to your program, sometimes problems don't become visible until you test over Samantha. You can test over Ad-Hoc Wifi networks (although using a router is preferred), so there's no excuse not to.

NOTE: There is an indication that in the 2011/12 season the software drivers will be updated to allow programs to be uploaded over Samantha. If that happens and it works well then there is little reason to use Bluetooth at all, even during program development.

 # 180. TIP: Obtain an older model WiFi router to test at your practice field, for example a wireless-G model, rather than fancy (and more expensive!) newer models.

The old models actually work more easily than the newer ones with the Samantha module. Some newer routers (or access points) cannot by default go slow enough to work with Samantha, this is not a problem with older models. The benefit of this is that older routers are actually cheaper. For example, we use a Linksys Wireless G router and it worked right out of the box. Another team called us up because they could not get their pricey extended range N router working. We advised they go buy a $50 G router instead, and they reported back to us later that it worked immediately with the cheaper router!

 # 181. TIP: Turn your laptop's WiFi off at all competitions

Having WiFi enabled on your laptop can interfere with the tournament's WiFi, even if you're not connected to any networks. Turn it off. You can't test over Samantha anyway at a tournament because you would have to overwrite the tournament's Wifi configuration on your Samantha Module.

Also, turning off Bluetooth on your cell phone, or any other device nearby is recommended. Bluetooth uses the same 2.4 GHz band as WiFi does, so Bluetooth devices can interfere with WiFi signals too.

 # 182. TIP: Turn off Bluetooth on your NXT brick before running over Samantha.

If you don't turn off Bluetooth on your brick before running a match, you can potentially be connected to your laptop and the Samantha FMS simultaneously. This can cause several problems.

A common symptom of this problem is that your Autonomous program moves the robot as soon as you run the program (before the match has started).

 # 183. TRICK: Monitor incoming joystick packets, if one doesn't come in for a long time then you have lost communication and need to shut down drive motors for safety.

It's not uncommon to drop the WiFi connection while driving the robot. When this happens, the default behavior is for the robot to keep doing what it was doing when the connection dropped. This can lead to a damaged robot if the bot rams into something while out of control.

You can stop the robot when communications are lagging by using code like this:

```
#include "JoystickDriver.c"

task main(){

  waitForStart();

  int lstmsg = 0;

  while (true){

    getJoystickSettings(joystick);
    if(lstmsg < ntotalMessageCount){
```

```
ClearTimer(T2);

      lstmsg = ntotalMessageCount;

      // New joystick messages have been received!
      // Set your drive motors based on user input
      // here.

    }else if (time1[T2] >200){

      // We have not received a packet in over two-tenths of a
      // second, which probably means communications have been
      // lost.
      // Put code to stop all drive motors to avoid
      // damaging the robot here.

      PlayImmediateTone(3000, 1); // play a warning tone

    }

    //Code here is run whether you have a connection or not

  }
}
```

The above code monitors the incoming joystick packets, and if it doesn't receive one for 200 milliseconds, it stops all drive motors and plays a tone.

Every time the NXT receives a packet, it increments a variable called ntotalMessageCount. In the example above, the code compares the current messagecount to the message count from the last time the loop executed. If there are new joystick messages, it resets timer 2. If timer 2 is ever more than 200milliseconds, that means no packets have been received for too long, so you should zero out your drive motors.

You should add your own code to stop the drive motors, stop any actuators based on 12v motors, or respond to userinput, where the comments suggest. You can put code outside the "if" statement, but keep in mind it will run whether or not there is a connection—so don't do anything potentially harmful to the robot.

The example works over both Bluetooth and WiFi, and should continue to work with any future wireless systems that FIRST uses. You can test this code by simulating a dropped connection by clicking "Pause Comm Link" in the RobotC Joystick Control dialog. Whenever communications are restored, the robot will begin responding again.

Part V: Electrical System

The electrical system can be a source of many problems, but by following a few simple tips you can prevent major difficulties from arising in the first place.

General Electrical System Tips

 # 184. TIP: Learn to solder.

Before the 2009/10 season, soldering was not permitted. As a result, the only way to connect leads to the motors connections was to clamp them together. This was not effective, and the leads would fall off constantly. There was also no good way to extend wires so they could be neatly routed around the robot. When soldering came into play, these problems became non-existent. Soldering is also useful for Prototyping Board work, extending wires, and many other purposes that will be explained elsewhere in this book.

Consult a mentor or online sources for proper soldering procedure, adult supervision is required, follow all safety rules, and always unplug the soldering iron immediately after finishing a task.

 # 185. TIP: A good place to get electronic items (including soldering kits) is www.elexp.com (Electronix Express).

A good soldering station is an important part of any workshop, and in FTC robotics it can be used to extend wires or solder motor wires. Electronix Express has low cost soldering kits, as well as other parts that may be useful for the ProtoBoard.

 # 186. TIP: Obtain an inexpensive multimeter so you can check voltage levels at the controllers.

A multimeter is a device that can be used to measure voltage, resistance, current, and sometimes other electrical values. You might think such a thing is expensive, but you can often find multimeters that will work just fine for FTC purposes for about $10 or even less on sale.

Frequently when you're debugging an electrical issue it's a good idea to just see if there is any voltage at all getting to the motor controllers. To do this, you simply select "voltage" mode on the multimeter's dial, then carefully place the probes on the battery (+) and (-) terminals on the motor controller in question. If you read between 13 and 14 volts there, then you know that the cables, battery, fuse, etc., are all in working order and you have a connection. If you read zero volts (or close to zero), you know some connection is loose, a cable is broken internally, the battery connector is bad, the battery fuse is blown, or some other problem is stopping a complete circuit from forming.

Batteries

187. TIP: *Use rechargeable batteries in the NXT Brick, not disposable batteries. Between rounds plug in your brick to keep it topped off.*

It is far simpler to use rechargeables than bring loads of throw-away batteries to the tournament. You can also charge the batteries while leaving your NXT brick connected to the field control system. Changing disposable batteries requires you to lose your connection and re-establish it, wasting time.

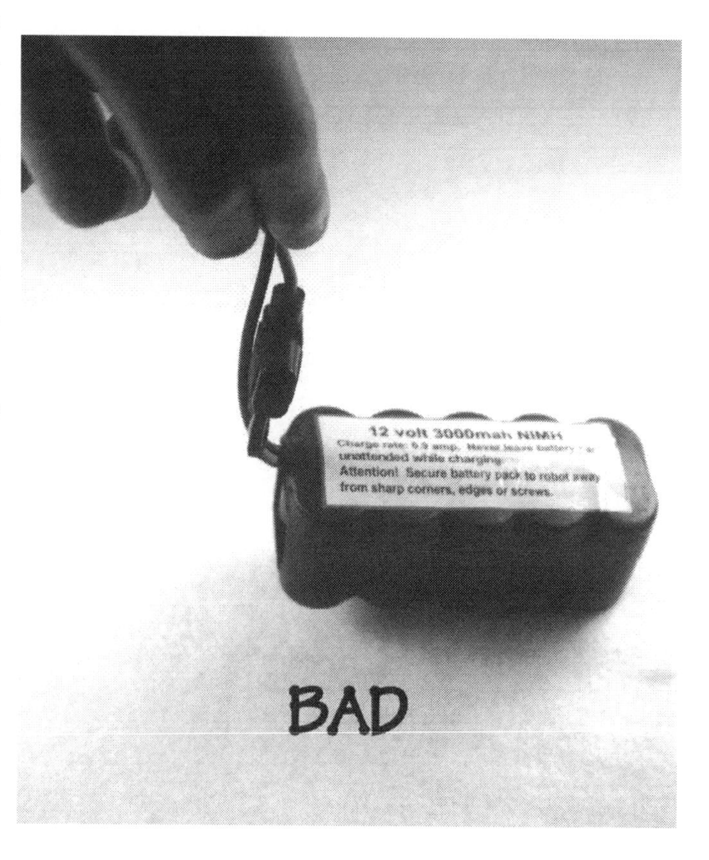

188. TRICK: *Never, ever pick up the 12v battery by its wires.*

Only pick up the battery by the case, not the wires. The wires eventually work themselves loose from the battery, or the connector on the end pulls loose making for a bad connection. The last thing you need after all your hard work is to lose a round because of a preventable battery failure.

Grill this into every team member! Whenever you see someone pick up the battery by the wires shout out, "Don't pull the wires!"

 # 189. TIP: Always charge the 12v batteries at 0.9 amps charging rate, never 1.8 amps.

The wires can actually melt off if you charge at the higher rate. It has happened to us. Starting in late 2010 new batteries carry this as a warning, but older batteries do not have the warning label. Regardless of whether you have new or old batteries, never charge any of them higher than 0.9 amps.

Whenever you plug a new battery in to be charged, visually confirm that the charging rate switch is set to 0.9 amps.

You may even want to put a piece of clear tape over the charging rate switch to make sure nobody accidentally switches it up to 1.8 amps (clear tape so you can see and visually confirm the setting).

 # 190. TIP: If you feel a battery while it's charging and it seems hot (not just slightly warm) to the touch, unplug it immediately.

There can be situations where a battery was nearly fully charged, and you plug it in again. The charger tends to power it for some minimum amount of time regardless of how charged it is, and that can cause it to overheat. It's normal for the battery to feel somewhat warm while charging, but if it's uncomfortably hot to the touch disconnect it immediately. Let it cool a while before using it. Check to make sure the charger is set to 0.9 amps.

 # 191. TIP: Never leave batteries charging unattended. At the end of each build session make absolutely sure all chargers are unplugged before you leave.

Batteries can overcharge and become damaged if left plugged in too long. Make absolutely certain all battery chargers are unplugged at the end of each build meeting, especially when the next meeting is a long time off (for example around holidays). Putting all your chargers on a single power strip and making sure it is switched off or unplugged at the end of each meeting is a convenient way to make sure they are all off.

 # 192. TIP: If a battery falls onto the floor or otherwise receives an impact, be extremely careful. Damage to a battery can cause it to short out and overheat. Notify an adult.

We found this out the hard way once when a battery was accidentally knocked off a table onto a concrete floor. Nobody paid much attention until someone said, "Do you smell something burning?" Smoke was pouring out of it! Now we usually charge our batteries on the floor under a table where they are out of the way. If they're already on the floor they can't get knocked down!

 # 193. TIP: Have a system to keep track of which batteries are fully charged.

A great technique that some teams use is to put a rubber band around fully charged batteries. When a battery is removed from the robot after use, the rubber band is also removed, showing that the battery has been used. When a charger light turns green, the fully charged battery is disconnected from the charger and the rubber band goes back on (and perhaps put in a box marked CHARGED for good measure). These techniques will prevent you from mixing up fully charged vs. partially uncharged batteries during a hectic tournament. Running your robot on a full charge is crucial for top performance during tele-op as well as consistent autonomous operation.

 # 194. TIP: Mark all batteries with your team name.

You can use an indelible marker, or print your team logo on self-stick labels and apply them (but do not cover the warning label). This goes for both 12v and NXT Brick rechargeable batteries. At tournaments items like batteries can get swapped by accident, or perhaps your alliance partner will need to borrow one from you. Keeping them marked is a good way to improve your odds of getting them back again!

 # 195. TIP: When you buy a new battery, mark it with the month and year you purchased it.

Batteries are fine for a couple of seasons, but they do lose their capacity over time; even the Nimh batteries used by Tetrix. For your most important rounds use your newer batteries. Use very old batteries (several seasons old) only for testing and demonstrations. Weaker batteries definitely do affect performance of the robot, speed, torque, lifting power, etc. Use an indelible marker on the body of the battery, or if you use stick-on labels with your team name you can mark on those labels.

 # 196. TIP: Keep spare fuses in your toolbox or emergency repair kit at tournaments.

If your NXT tells you the external 12v battery reads zero, this typically means one of four things:

1. The kill switch has been accidentally left in the OFF position.
2. The wires or connectors leading from the battery are either broken or otherwise not connected.
3. Your NXT brick needs to be rebooted.
4. The fuse is blown.

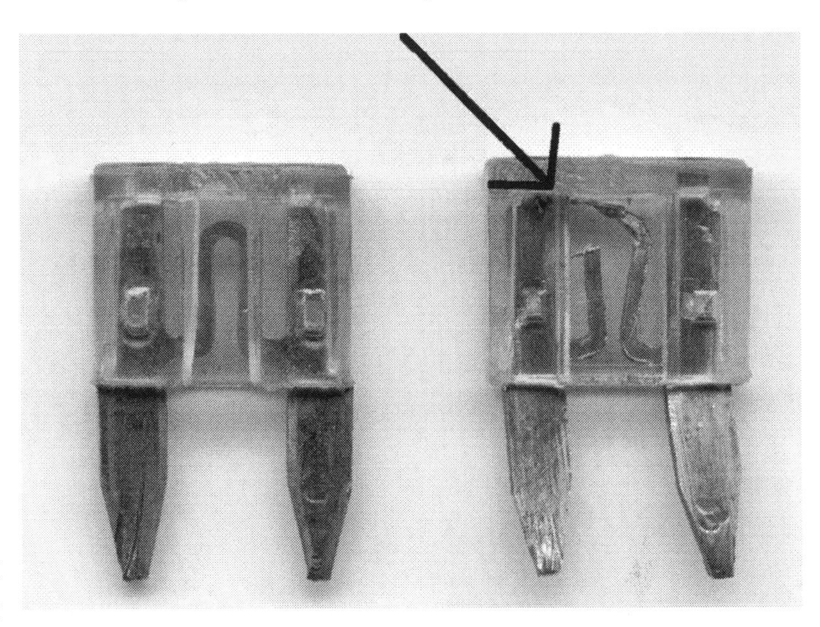

The photo shows a good fuse on the left and a burned out fuse on the right. Notice the telltale gap caused by too much current actually melting away part of the metal conductor (arrow). You may also sometimes see black scorch marks.

Fuses usually blow because there is a short circuit somewhere in your wiring. In other words, the red (+) wire has come in contact with the black (-) wire, or both have contacted a conducting metal surface.

Keeping a set of fuses in your toolkit can help you quickly salvage this situation, assuming you have found the source of your short circuit. If you replace a burned out fuse without having fixed the underlying cause of the short circuit, it will simply blow again. Trace your wiring to try to find the problem, look for frayed wires, wires not totally inserted into motor controller slots, motor connections that are not insulated or that contact the motor body, etc.

Fuses can be purchased where automotive supplies are sold. Take an old one with you to make sure you match the exact type and amperage. Never use the wrong kind of fuse, and **never under any circumstances use anything other than the cor-**

rect fuse to bridge the connection, as the batteries may overheat and even catch fire if too much current is drawn.

NXT Brick

 # 197. TRICK: Every time you flash the NXT Brick BIOS, you must remember to reset the inactivity timer so the brick never turns itself off.

Failing to do this may cause you to lose your connection while you are queued for an important round or worse yet, your brick might decide to go to sleep right as the round starts! You don't need that kind of stress. The inactivity timer is in the NXT firmware options menu.

 # 198. TIP: Put self-stick labels around the ends of connectors that plug into the brick to make it easy to match up ports with the right connectors.

Number them with an indelible marker according to the port number (1, 2, 3, 4) or the motor letter (A, B, C).

 # 199. TIP: Write your team number on the NXT Brick using an indelible marker.

Especially if you have a backup brick, you want to make sure you're using the same one that has all your latest programs and you've practiced with.

 # 200. TIP: Reboot your NXT brick between every round of play.

Bricks that run for long periods of time sometimes crash unexpectedly. Like any computer, rebooting "cleans up" memory and reduces that kind of problem. We recommend rebooting the brick between every match. Remove the battery, count to 10, and then replace it.

 # 201. TIP: Don't let the NXT Brick battery fall below 7.9 volts.

At about 7.7 volts you can lose your Bluetooth connection, for example. If you are using some NXT motors which draw power from the NXT brick, you may want to even change batteries or plug in at 8.0. Plugging in the NXT rechargeable battery between rounds to keep it topped off is never a bad idea.

 # 202. TIP: Don't let the 12v battery fall much below 14.0 volts at the start of any round.

Although the robot will probably still function down to 13.5 volts or even a little lower, you want to have the most speed and power possible in your round. Lower voltages especially can affect autonomous mode. We typically swap batteries between each and every round of play when possible, but at most use the same battery for 2 rounds of play. If you are in the elimination rounds you need to be very careful to keep your batteries charging when not in use.

Wire and Cable Management

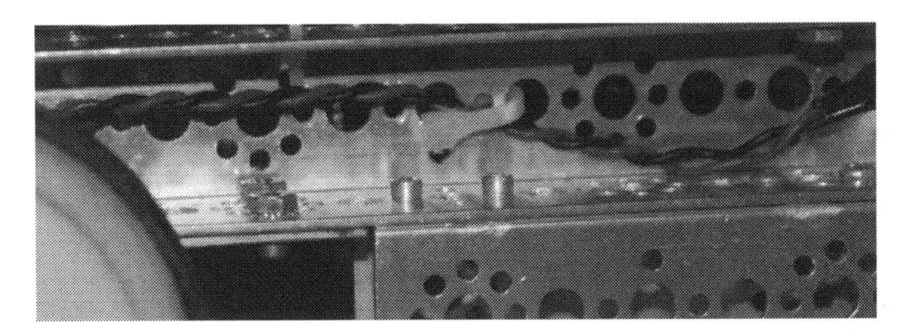

 # 203. TIP: Try to route wires through channels where possible.

It keeps them safe from entanglement and neat looking.

 # 204. TIP: Make sure the kill switch is mounted in a place that makes it highly unlikely for a collision to accidentally trip the switch.

We have actually seen robots turned off by a collision with an opponent! Mounting the switch on top of the robot, or recessing it back an inch or two into the body of the robot, or even mounting it inside a channel, can provide enough protection to make an accidental turn-off very unlikely.

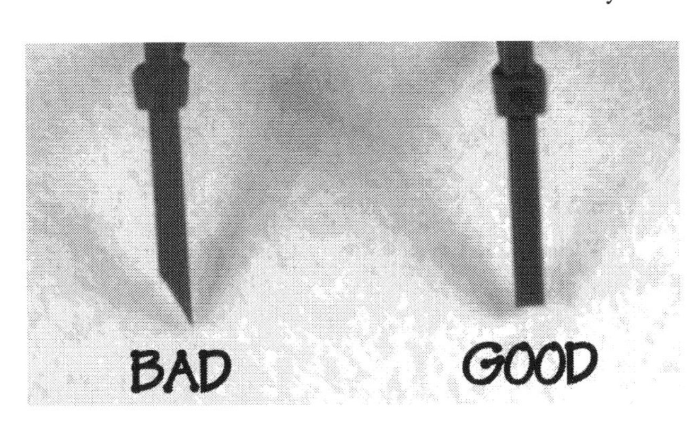

In the photograph, the kill switch is mounted so it faces upward—this makes it unlikely for a collision to accidentally shut off the robot (as opposed to facing the switch outward toward an oncoming opponent or field element). Just make sure it is still easy to see and is accessible in case the referee has to disable your robot for safety reasons during a match.

 # 205. TIP: Never trim cable ties at an angle, always trim them straight across.

If you trim a cable tie at an angle it can form a very sharp knife-like structure that slices through skin even at the merest touch. Cut straight across, and then carefully check for sharpness with your thumb, if it's still sharp hit it with some sandpaper a couple of times to dull it.

 # 206. TIP: Solder the 12v motor connections.

Starting with the 2009/10 season you are allowed to solder the leads onto the 12v motor. We found this to be a much better way of ensuring connections than the prior season. There is also a plastic motor clamp available, but we have not tested it extensively. We *do* know that in over 60 real rounds of play in tournaments and scrimmages, and many dozens of practice rounds, we never had a single motor fail because a soldered lead came loose.

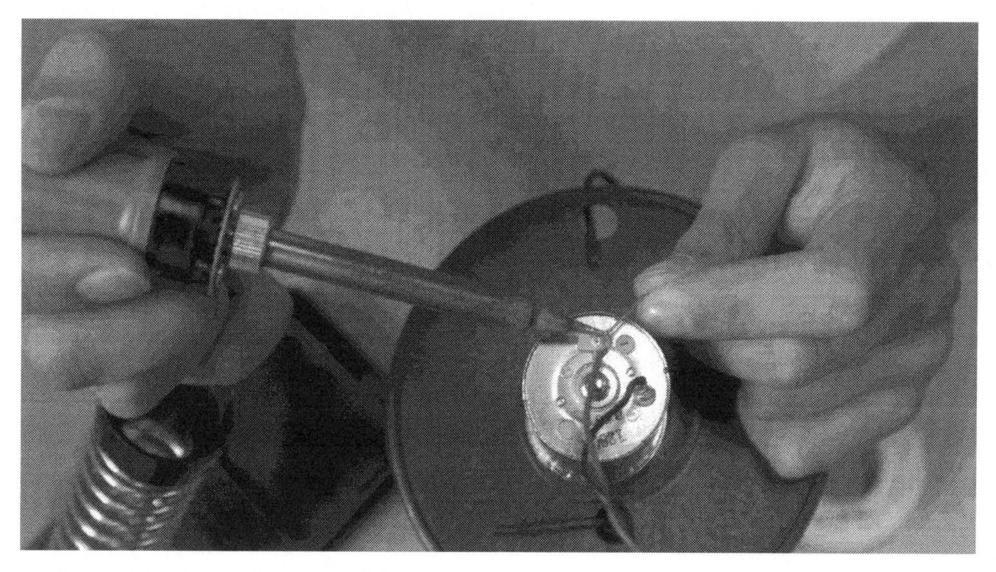

Important note: if you do not have an adult mentor who knows about soldering to instruct you, then you are probably better off using the snap on connectors. Heating the motor connector for more than 10 seconds, or with too hot an iron, may damage the motor internally, causing it to work poorly or not at all. In the 2010/11 season a new "fused" connector is also available that may prevent motor burn outs. We do not have enough experience with these yet to make a recommendation.

Make sure you insulate with electrical tape or heat shrink tubing after soldering and inspecting.

 # 207. TIP: If soldering, inspect 12v motor solder joints using an eye-loupe; make sure the solder does not bridge to the motor body.

An inexpensive eye-loupe or magnifying glass can help you inspect solder joints to ensure they are not contacting the metal body of the motor.

 # 208. TIP: If soldering, have a second student check that the red wire is (+) and black wire (-) as quality assurance after soldering.

Having two people check the connections increases the likelihood of it being correct. Near the motor connection joint (on the motor) is either a (+) or (–) symbol. Make sure the leads are connected accordingly. Red wire should be used for (+) and black for (-).

 # 209. TIP: Use soldering to extend wires as allowed by the rules so you can route them neatly.

Soldering extensions onto the wires was allowed in the 2009/10 season. Make sure you use the proper gauge wire. For motor wires, use 22 gauge, and for battery wires use 16 gauge. Consult electronics tutorials on the web for proper soldering technique. Adult supervision and appropriate training for students is a must.

You should make the wires longer than necessary, and then later trim them to make them look neater. Always leave a few inches of slack in case you need to fix a break or reroute.

Always use electrical tape or heat shrink tubing to electrically insulate soldered extensions.

Motor and Servo Controllers

 # 210. TIP: Make sure there is no bare copper showing on any wires leading into a motor or servo controller, not even a single strand of threaded wire should be visible.

The controllers can be damaged (or the battery fuse blown) by a simple arc across the contacts. If you can't shove the wire all the way in to the connection receptacle so no bare copper shows, then trim the wire a little shorter and try again. We have seen robots with actual burn marks around the contacts because of shorting caused by sloppy connections into the controllers. This can cause a catastrophic failure of your robot as well as burning out an expensive controller. If you are using stranded wire, tinning the end of the wire using a soldering iron will prevent little strands from peeling off that are hard to press into the connection hole. But be careful not to use too much solder which could make the wire too thick to fit in the controller connection hole.

In the photo, the two wires on the left are not totally inserted into the controller slot, you can see bare copper wire exposed (circled). The two wires on the right are fully inserted, you can only see insulating plastic wire covering, not bare copper. If you use stranded wire that is not tinned with solder, be sure that not even a single strand is exposed, that's all it takes to arc across the contacts and cause serious trouble.

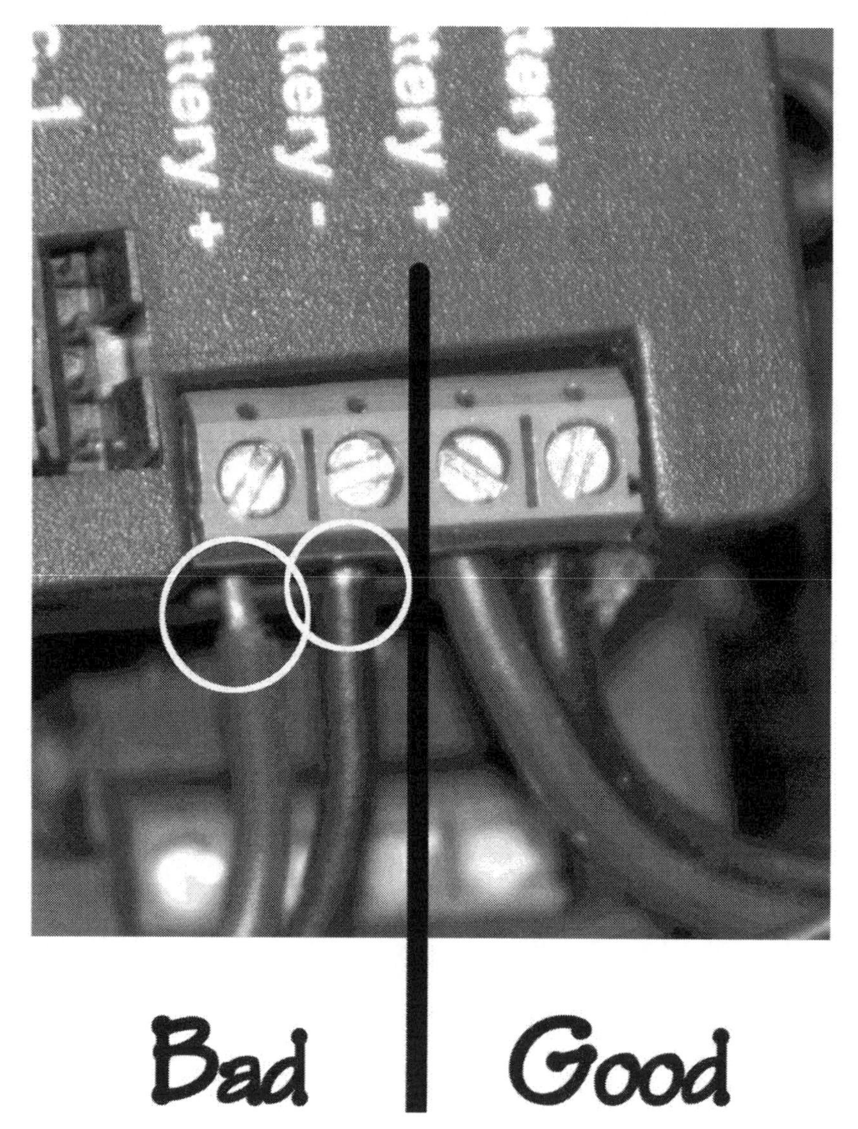

#211. TIP: *If your motor seems to be behaving oddly, make sure nothing has fallen into unused connection ports on a controller.*

In the 2008/09 season one of our robots was disabled for the first four rounds of a tournament, and the problem turned out to be that a nut had fallen into one of the connector ports on a motor controller. As the robot moved around the nut would jiggle and sometimes short out the connections, sending random data to the NXT brick, causing very odd problems. By following other tips in this book (i.e. keep the controllers on the bottom of the robot protected by plastic) you can make this very unlikely to happen. Another technique would be to plug short unused cables into the unused ports so nothing can ever go into them.

 # 212. TIP: Daisy chain your controllers to conserve sensor ports.

You can connect several motor and servo controllers in series so that you use only a single NXT sensor port. If you do this, be sure the very first controller connected to the NXT is a motor controller and not a servo controller. The motor controller has to be first in the chain so that your brick can get battery status information.

Part VI: Programming and Sensors

Programming the robot can be challenging, and there are many ways to go wrong along the way. The tips in this section will help you to avoid the most common mistakes and hopefully will smooth your way toward solid programs both for autonomous and tele-op, as well as standalone programs for use in testing.

Sensors are also covered in this section because use of sensors always involves programming. (The exception to this is that sensor mounting techniques are covered in the section on Building.)

Code Samples

We have posted example code, free for anyone to use, at:

```
http://code.google.com/p/libftc-util/
```

Simply go to that location, click on "Downloads" and the samples are in a zip file.

General Programming Tips

213. STRATEGY: *Always backup your programs before the start of each programming session by copying the entire folder containing your programs to another folder that contains the current date.*

Failure to do this may lead to pain and heartache if you accidentally break your program and have failed to save the last good version.

214. STRATEGY: *In addition to the above, back up your programs to external media such as a USB flash drive or memory card.*

In case your laptop hard drive fails, or your laptop is lost or stolen, you want the programs stored somewhere that is not physically connected to the laptop.

 # 215. STRATEGY: Furthermore, burn your programs periodically to a CDROM or DVDROM. Include installation files for the field control system and RobotC programming install. Bring this to the tournament.

Creating a CD or DVD that contains all your programs and all your installation files is another way to recover if something goes wrong at tournaments. Storing these periodically in a location away from your laptop also is a further disaster recovery precaution.

 # 216. TRICK: Use a source code control system.

Using a source code control system, such as Subversion or Git, allows for many time saving opportunities. If there is only one software developer for your team, then you can use a system like Git. Git is a distributed revision control system. This means that it does not depend on a network resource, which saves you the money and time to obtain one. It also controls your changes, acting as an ultimate undo button. At any time, you can reverse your changes to a prior version, which you make periodically. Revisions can be made at the end of every meeting, or after every significant change. A centralized revision control system, like Apache Subversion or Mercurial, is best when there are many developers. This system stores the revisions in one remote server, and each developer "checks out" a copy of the code. Then, they make the day's changes, and commit them to the server. This system keeps everyone synchronized, and allows for conflict resolution if two people change the same area of code.

For more information, see Appendix B: Setting Up Source Code Control Systems.

 # 217. STRATEGY: Comment your code right from the beginning. It will aid in debugging.

All programming, whether text based or not, has the ability to be commented. Comments are human readable pieces of text that communicate to the programmer what they were thinking when they wrote the piece of code. The primary feature of comments is to let other programmers understand your code when you are not present. Properly commented code should be understandable to any programmer familiar with the programming language. Also, comments are useful to remind you what you were thinking when you wrote the code; there are different ways to interpret the same code block, and there are many ways to solve a single problem. Your comments should easily communicate the intent and method of the code.

218. STRATEGY: *Variable names are crucial for code readability. Avoid cryptic names.*

When creating variable names, programmers will often pick a name that makes sense right then, but when they come back to it days or weeks later it no longer makes sense. Sometimes programmers will even use internal names of structures which they should rename, like motors and sensors. A name like "motorLeftFront" is much more descriptive than "motorLF" or "mtr_S1_C1_1." A little extra typing goes a long way for readability. Also, if there are many variables that belong to the same physical structure or concept, like the drive train, give them all the same prefix, or place them in a "struct."

219. STRATEGY: *Program to be fail safe. If a sensor fails, have the program detect this and at least do something reasonable.*

You never know what's going to happen during a tournament. Hardware failures are common. Sometimes a sensor could become unplugged or fail for some other reason. In many cases, such a sensor will return an error value, such as -1, if it cannot be read by the program. In cases like this, you should detect such failures and then do something reasonable.

For example, if during your autonomous program you detect a sensor is dead, you might want to just do nothing more at that point, simply stop. That could save you from wasting pre-loaded game scoring elements, for example. If you want to be more sophisticated, you could choose a different strategy in case of failure that does not depend on the bad sensor. For example, if a light sensor that is being used to follow lines in autonomous fails, you could revert to a dead reckoning version of the program. It won't be as accurate as following the line, but you might score more than simply stopping the program.

If you use such strategies, it is important to test your code both ways. For example, unplug the sensor to make sure the fallback code really works, then plug it back in and try again to test both paths through the program. As your program becomes more sophisticated, testing becomes critical to ensure all of it is working the way you intended.

Laptop, Field Control System, Drivers, and Other Software

 # 220. TIP: About a week before any scrimmage or tournament check to see if the Field Control System or your software programming system (RobotC, NXT-G, or Labview) has been updated. Make sure you practice using the same version as you will run with at the competition.

Especially early in the season there are likely to be several updates to the software. You may be rejected during software inspection if you are not running the very latest version of the Field Control System.

Your programming system may have had critical bugs fixed as well and should be updated, but it is also important that your final practice before any competition should be run using the latest software to ensure that the upgrade does not introduce any new problems.

If you are closer than about a week to your next competition and won't have time to do a full practice run, don't do any software upgrade.

 # 221. TRICK: In the RobotC development system, keep the menu level in either "Expert" or "Super User" mode.

In RobotC, if you are in the "Basic" menu level, many options are hidden, several of which are desirable. To show these options, set the menu level at "Expert" or "Super User" by going to the menu: Window/Menu Level/Expert, or Window/Menu Level/Super User.

 # 222. TRICK: In the RobotC development environment, use options to show additional sensor types.

In RobotC, many sensor options in the Motors and Sensors Setup screen are hidden by default. To show them, put RobotC into Expert or Super User mode (see the TRICK above), and open the Motors and Sensors Setup screen. Go into the tab named "Device Mgmt NXT", and under "Allowable 3rd Party SensorTypes", check the boxes of any sensor types you want RobotC to show in the sensors setup screen.

Debugging Strategies

 # 223. TIP: Learn how to use RobotC's debugging windows.

RobotC gives you ways to monitor variable values right on your laptop console, and this can be very helpful for debugging program errors. Sometimes there can be some "lag" between updates, however, and in real-time situations like debugging an autonomous program you may need to be looking at the robot and not the computer console to determine where problems are, so see the next few tricks for alternative ways to get feedback that do not involve taking your eyes off the robot.

 # 224. TRICK: Use sounds for debugging.

Insert various tones or sound effects in your code for debugging purposes. By playing different pitched sounds at certain critical points in the code, you can figure out which part of the code is running and observe if the robot is doing what it's supposed to. This is especially helpful in debugging a piece of code with many loops and functions, or when a conditional statement may or may not execute. To play a tone using RobotC use:

```
PlayTone(500, 10); // play a tone of 500 hertz for 100 milliseconds
```

 # 225. TRICK: Use time delays for debugging.

By putting time delays between different parts of code early on, it is easier to determine what is happening. This is especially true if you have a complex autonomous mode with many robot actions. Putting a second or two between each action will slow things down and show you where you might be going wrong. The time delays should be removed for an actual tournament, as you do not want to waste time. One way to do this is to put the time delays inside conditional statements that only run if a global "debug" variable is set to 1. In RobotC you could use something like this:

```
int debug = 1;  // change this to 0 to turn off debugging

.
.
.

MoveForward(900);    // move forward for 900 msec

if (debug) { wait1Msec(1000); }
```

 # 226. TRICK: Use LEDs or servo motors for debugging.

Another method of debugging is to use Lego LEDs or even servo motors to give a visual indicator. For example, when a certain condition is detected such as an ultrasonic sensor detecting an object, have the LED light up as an indicator. Using a servo motor with a piece of metal attached to serve as an indicator pointer you could even represent the distance that the Ultrasonic sensor is reading. For example turn the servo half a degree for each centimeter of distance. Taping a piece of cardboard behind the pointer could even let you read off a fairly accurate scale. If such an indication were required during the real game at a tournament you would of course have to construct such a scale from legal materials.

Autonomous Programming

In our experience nothing separates champion teams from others more obviously than the sophistication of their autonomous mode. There is a simple reason for this: autonomous mode programming is *very hard* if you don't have prior experience to help guide you. But if you do develop a good arsenal of autonomous programs, you will have an insurmountable advantage over all but a handful of teams. Your chances of becoming a champion team increase dramatically if you master autonomous mode.

At the Regional Championship tournaments we've been to, and that includes the New Jersey, New York City, and Pennsylvania Championships, most teams do not have any autonomous mode at all. In many rounds of play, none of the four robots even attempt to move during autonomous. In the majority of rounds you will only see very simple autonomous modes that perform low scoring tasks like opening dispenser chutes for a few points.

In recent Regional Championships we've been to, no more than about ten percent of the teams will have an autonomous mode that can score a game-changing number of points.

You really want to be one of those teams!

The benefits of a truly high scoring autonomous mode cannot be understated. In many of our regional championship rounds, we had already scored enough points to win by the end of the 30 second autonomous period. This allowed us to win rounds sometimes even when our partner robot was totally inoperable, or even failed to show up at all! It also allowed us in some cases to run up a good score early and then score on our opponent's behalf to gain ranking points and an edge in tie-breaks. The benefits just pile up exponentially.

This chapter is going to try to help you overcome the daunting aspects of programming autonomous by steering you away from some common mistakes that prevent most teams from creating a great autonomous program.

Types of Autonomous Mode

There are several different general types of autonomous strategy. A good team will develop several, or even all, of these, for use in different situations:

- Defensive Autonomous
  A defensive autonomous will try to stop the other team's autonomous from achieving its goal. For example, if you believe your opponent will drive to a certain position on the field to score, your autonomous could drive to that same place and either block or ram them to prevent scoring.
- Counter-Defensive Autonomous
  You don't see too many of these come into play until elimination rounds or the World Championships. A counter-defensive autonomous tries to block an opponent who is going to try to stop your partner robot's autonomous by ramming them. In other words, it's trying to ram the rammer! We used this strategy at the Pennsylvania Championships and in the World Championships several times. We have yet to see a Counter-Counter Defensive Autonomous, but we're sure that will happen eventually…
- Low Scoring Autonomous
  A low scoring autonomous will attempt one of the simpler autonomous tasks defined by the game. In recent years you could score 5 to 10 points by opening a dispenser containing game elements, for example.
- High Scoring Autonomous
  A High Scoring Autonomous will attempt to score multiple game elements in a high scoring goal. This is typically much more challenging than defensive or low scoring autonomous modes. In the 2008/09 season it was necessary to dispense game elements first to score significant points, adding to the challenge. In the 2008/09 season the best teams could score 70 to 120 points, and occasionally more, during autonomous. In the 2009/10 season the best teams were often scoring 60 to 80 points, and we heard of one team able to score 190 by first dispensing (but they could not achieve this consistently). During regional championships that we attended in 2009/10, scoring 50 points in autonomous was almost always enough to seal the game.
- Harvesting Autonomous
  A harvesting autonomous attempts to dispense and harvest as many game elements as possible. This denies those game elements to opponents during tele-op. Sometimes such programs will first harvest from one position, then

may drive to a second position and park so that during tele-op the drivers can immediately harvest a second time. This can lead to an overwhelming advantage. If you can control half or more of the game elements before tele-op even begins, your opponents will in effect starve—they will have few or no ways to score.

- Combination Autonomous
  This type of autonomous combines two or more of the other kinds. For example, first run a harvesting segment to dispense and harvest, then try to score them. Or, first use game elements the robot starts with in a High Scoring mode, then after that go ram the opponent defensively to interfere with their plans, or travel over to a dispenser so the drivers can immediately harvest more when tele-op starts. The possibilities are really endless.

For your first scrimmage you may not have time to develop anything other than a simple low scoring autonomous. There is not much reason to develop a defensive autonomous for an early scrimmage because most other teams won't have anything to defend against, but if you know you have a tough competitor in your area that is likely to have an autonomous early in the season that could be an exception that warrants work on defense early.

By the time of your first Championship tournament, your goal should be to have a consistent High Scoring Autonomous that works from at least one field position, a Low Scoring Autonomous that works from the other field position, plus perhaps a Defensive autonomous. That gives you flexibility when planning strategy with your partner robot.

General Concepts

 # 227. STRATEGY: We recommend using RobotC as your programming environment.

In our opinion, it is more difficult (but not impossible) to create a competitive, complex autonomous programs in anything other than RobotC. Labview and NXT-G become more difficult to debug as the number of blocks and connections increase, and sophisticated autonomous programs will have many, many sequences of steps. Also, RobotC can compile and upload a new program in just a few seconds, far faster than the other environments in our experience. (Although we are told Labview's new 2011/12 version has made improvements in this area.)

Compilation speed becomes important when you are repeatedly running tests for your autonomous program and keep modifying values or trying new things. If you have to wait upwards of a full minute for each test as opposed to just a few seconds, that time adds up very quickly. If you can get thirty test runs done in an hour you're going to be able to develop a much better program than if you can only complete ten.

Labview does have the advantage of being generally easier for novices to learn, and that may be the most important consideration for teams that do not have experienced programmers, or who lack mentors with programming experience to teach students RobotC.

Of course it's possible that the other environments will improve and change this calculation as time goes on, and certainly we have found issues with RobotC. But as of this writing, RobotC's advantages appear to be significant for developing sophisticated autonomous mode programs.

 # 228. TIP: Use the code templates.

Yes, you must use the code templates provided on the FIRST web site to create your programs using RobotC or other environments. Not doing so can lead to problems, such as the robot running the program when it shouldn't. Also, if the software inspector at a competition discovers you aren't using the code templates like you should be, you may not be allowed to run your robot.

 # 229. SECRET: By the time autonomous programming begins, structural changes to the robot hardware must be complete, especially anything that has to do with the wheels and drive train.

Especially when an autonomous program is using one or more "dead reckoning" sequences, any change at all to the structure of the robot could cause the program to lose accuracy. This is obviously true if you change the drive train (wheel position, gearing, etc.) but it is also true that even redistributing the weight of the robot by modest amounts can change how far the robot turns and other factors. Simply moving the battery from one place to another could throw off movements enough to cause the robot not to score. So it is best to stop virtually all hardware changes as soon as autonomous programming starts, especially if you are getting near a scrimmage or tournament. After any significant hardware change later in the season, the autonomous programs must be tested and re-adjusted if necessary.

 # 230. TIP: Allow some "settling time" after a motion or turn.

When the NXT "brain" tells the motor controllers to make actions, it may have to dispatch commands to multiple controllers, which takes time. So, not all motors may start at the exact same instant. Also, your robot has mass, and should have a lot of it.

That mass will not stop instantly. You should allow about 100 milliseconds (0.1 sec) to let the robot finish all its movement actions and for the robot to physically settle on the floor. This is especially true if gearing is involved. By pausing for a moment, you ensure that your next turn or movement will continue after the robot is fully stopped and settled, as opposed to starting another motion while the robot may still be rocking or slightly moving and may not have full contact with the floor. While time is important, repeatability will be enhanced if the robot is in a settled state when each new movement begins.

 # 231. TRICK: Check both the NXT battery and 12v external battery when the autonomous first starts. If levels are too low play a distinctive warning siren.

In RobotC you can check the 12 volt battery level by referencing the variable nExternalBatteryAvg, which is in units of millivolts. For example, if this variable holds a value less than 14000 that means your battery is under 14 volts. The NXT battery level can be tested by referencing the variable nAvgBatteryLevel, and it is also in millivolts. We use a distinctive "warning siren" sound effect to signal that the battery is too low the moment the autonomous program is started. The lower the battery is, the more times we play the siren to catch the operator's attention. This can save you from losing a round by forgetting to turn on the "kill switch," or by accidentally using a battery that is too low or using a battery with a blown fuse. Example code:

```
if (nAvgBatteryLevel < 8000) {     // brick under 8.0 volts
  PlaySoundFile("NxtBatteryLow.rso");
}
if (nExternalBatteryAvg < 14000) {     // main battery under 14.0
  int i;
  // play a warning siren once for
  // every tenth of a volt too low
  for (i = 0; i < (14000 - nExternalBatteryAvg)/100); i++) {
    PlaySoundFile("ExternalBatteryLow.rso");
  }
}
```

 # 232. TRICK: Test your autonomous mode with at least a 14.0 volt charge on the external battery.

Autonomous will work at its best with a full charge. Testing with at least a 14 volt charge will increase consistency in the autonomous working as should, provided that all matches are entered into with a full or very nearly full battery.

 # 233. TRICK: Track how well your autonomous works at different battery levels so there are no surprises.

Battery level can have a profound effect on the repeatability and accuracy of your autonomous mode. This is especially true for dead reckoning maneuvers. If you are trying to perform more than one turn using dead reckoning in a movement sequence, even a tenth of a volt can be enough to destroy your accuracy. A single turn provides a little more room for error, but we find that two tenths of a volt starts to make a significant difference.

Sensors reduce the battery dependency but do not entirely eliminate it. It is good to know what effect different battery levels will have on the effectiveness of your autonomous mode, so run some tests at different battery levels and track the results.

 # 234. TRICK: Always set up the robot exactly the same way for autonomous, regardless of field position.

Unless there is some compelling reason otherwise, you should try to create your autonomous programs assuming that the robot will be set up precisely the same way in the starting box, no matter which field position or autonomous mode you are running. For example, the robot is always oriented with the harvester facing the center of the field, the right side wheels are lined up with the starting box's tape outline, and the back of the robot is touching and square with the wall. The precise setup will depend on how the game works, of course, but what is important is consistency. This reduces operator error during setup. If you have different ways of setting up the robot depending on the field position or autonomous mode, you are asking for trouble.

Of course, if one of your autonomous modes is wildly different than the others you may need to make an exception to this rule so that the program does not need to make unnecessary turns, but start out by trying to design a setup that can be common if possible.

 # 235. TRICK: Plan on programming autonomous to work from both field positions.

While for early scrimmages or even an early regional qualifier you may not have time to get your autonomous working from both field positions, you should still plan from the start to get both sides working. There will most likely be a time where a partner robot can only run autonomous from one field position. It is almost always most beneficial to have two autonomous robots running on a team, so if this situation occurs, the robot with two autonomous modes can still run from the other position.

This is especially important during elimination rounds when you are more likely to pair with a partner who has their own autonomous mode. If both of your robots can only work from the same field position, that would mean one of you would have to disable autonomous and just sit there rather than accomplishing something useful.

 # 236. TRICK: Use NXT buttons to select which field position the robot starts on instead of having two different programs.

You could create different autonomous programs for different situations, such as running from the left or right field positions. But instead, it is better to have a single program that starts up and reads key presses from the NXT brick itself to select those kinds of options. By using the NXT buttons, there is less chance that the wrong autonomous is run, as you are always running the same program but simply selecting different options. Usually in an autonomous program, there is a part of the code that will be common regardless of which field position the robot starts in. By using the NXT buttons, you only need one program, and the logic that runs in both does not have to be repeated, therefore saving valuable space on the NXT and allowing you to maintain a single program.

 # 237. TRICK: Another way to select field position is with a touch sensor.

Instead of using the method outlined in the prior TRICK, some teams use a touch sensor to select left or right field position options during autonomous. A simple technique is to use a rubber band to hold the sensor in the "touched" state for one field position, then release it for the other. The advantage of this method is that it can be hard to read the NXT screen under some lighting conditions, and the rubber band is easy to see, confirming you've got the right option selected. The disadvantage of this method is that you use up a sensor port. (Also, while in recent years rubber bands were allowed materials, before you use this strategy make sure the new season game rules allow rubber bands.)

 # 238. TRICK: Another way to select field position is with a prototype board switch.

Along the same lines as the last couple of TRICKs, if you have many different options to choose in your autonomous it might make sense to build a set of switches interfaced to the Hightechnic Prototyping Board. For example, in the 2009/10 season we had 14 different autonomous modes which were all variations on about four different basic modes. We used a potentiometer to select a "major" mode, a slider switch to select left or right field position, and another slider switch to select whether or not balls stored in the harvester should be shot. (On some fields the goals spun very easily and it was best to hold the extra balls for tele-op, on others the goals did not spin much and it was better to try to score them during autonomous for a higher point total.) See Part VII: The HiTechnic Prototyping Board for how to interface switches and use a potentiometer to select from several autonomous choices.

 # 239. TRICK: In some games you can select field position automatically using other kinds of sensors.

In some cases you may not need to tell the robot which side of the field it is starting on, instead sensors can be used to detect position automatically. A good example of this was the 2008/09 season. Our robot followed the wall to a dispenser which rose up above the level of the wall. The robot had an ultrasonic range finder mounted on both its left and right sides. When we set up the robot in autonomous, we always pointed it toward the dispenser. Now, as it advanced forward, either the left or right ultrasonic would detect the dispenser, telling us whether we were on the left or right field position automatically! There was no need to use switches or buttons to tell the robot where it was on the field, it could tell by sensing which of the two ultrasonic sensors triggered first.

 # 240. TRICK: Use sounds to confirm field position when autonomous starts.

By having sounds in the initialization procedure, you can confirm which field position the program thinks the robot is starting from. You could use different tones, such as high pitch to mean "left" and low pitch to mean "right," or you can get fancy by using custom RSO sound files created by the program wav2rso that actually say the words "Left" or "Right." See Appendix D for more information on creating your own custom sound files for the NXT.

 # 241. TIP: Time matters in autonomous. Faster is better as long as it does not sacrifice too much accuracy.

There is frequently a balance between speed and accuracy. Some of the other tips in this book, such as to put short time delays after each autonomous motion, reduce speed but may increase accuracy. When you are going up against robots that do not have their own autonomous mode speed does not matter much, but against tougher competition it often matters which robot gets into scoring position first; it is a race. You may also face robots that have defensive autonomous modes that attempt to ram your robot before it scores—so the faster you score the better.

Some techniques for moving into position are faster than others. For example, line following is generally slower than dead reckoning. This tip simply asks you to consider carefully this speed versus accuracy trade-off when designing your autonomous strategy, and try to make your autonomous mode as fast as it can be while still achieving the required accuracy.

 # 242. SECRET: Be ready to adjust your autonomous even during a tournament based on competitor or partner robot capabilities.

The best teams will create brand new autonomous programs even while they compete in a tournament in order to counter strategies they see their opponents employing, or to accommodate a partner robot's autonomous strategy.

For example, in the 2010/11 season a small number of teams were able to cross the field and dispense a "doubler baton." But in order to counter this strategy, one member of an alliance merely had to drive forward about 5 feet during autonomous to block the dispenser. A good team, even if they had not thought to program that ahead of time, would be expected to get that working even during a tournament to stop that opponent. In other cases, you may simply need to insert a time delay to allow your partner robot to cross in front of your robot before you start moving in order to avoid a collision during autonomous with your partner.

Many other cases like this will come up frequently during tournaments and you must be prepared to make new programs as needed to win games.

One strategy to make this easier is to code up small snippets of code that you can simply copy and paste together to form new autonomous programs or adjust existing ones.

Whenever making such changes, proper backup and naming of your program code is essential. Be certain that the working code you walked into the tournament with always remains untouched, allowing you to revert back to a known good state when needed. Always name new versions created at the tournament very clearly to avoid confusion.

Whenever possible, test your new program on a practice field. If that is not possible, for simple actions like "drive forward 5 feet" you can perhaps run a test right in your pit area, or find a clear hallway somewhere to test. Just warn people around you that you are about to run the robot, and be ready to hit the kill switch if things go wrong!

Sensors

 # 243. TIP: Use timers to avoid burning out motors whose limits are detected by sensors.

Because there is no driver during autonomous mode it is very easy for the robot to damage itself. For example, a stalled 12v motor will start sustaining heat damage in about 5 seconds, and in 10 to 15 seconds it will literally go up in smoke. Proper programming can reduce the chances of this kind of mishap.

For example, let's say you have an actuator of some kind being driven by a 12v motor. Further, let's say that you are using a touch sensor to detect when that actuator reaches its full extension. Now, let's say your autonomous program is written such that at some point you attempt to move that actuator, and you continue to move it until the touch sensor triggers. Something like this:

```
Move_Actuator(100);  // turn on motor at 100% power forward

while (SensorValue(actuatorTouch) == 0) {
      wait1Msec(100);   // wait until you see the touch
}
Move_Actuator(0);    // turn off the actuator motor
```

This is a very poor way to proceed! The reason is, if something prevents the actuator from reaching the touch sensor, such as an opponent ramming you, getting tangled up in a field element, or the touch sensor failing, then the motor is going to stall and your program will happily continue to power it until smoke pours out and it's destroyed. Then, not only does your autonomous program fail, but your drivers will not have the use of that actuator during tele-op, which could be disastrous if it serves a critical function like scoring or harvesting.

In this case, you really want to implement an absolute time limit to save your motor, like this:

```
Move_Actuator(100);  // turn on motor at 100% power forward

ClearTimer(T1);       // Use timer T1 to limit actuator for safety
while (SensorValue(actuatorTouch) == 0 && time1[T1] < 2000) {
      wait1Msec(100);   // wait until you see the touch sensor
}
Move_Actuator(0);     // turn off the actuator motor
```

In this new code, a timer is cleared before the while loop, and the while loop exits either if the touch sensor triggers or when the timer value exceeds 2000 milliseconds (2 seconds). This means you will not continue to power the 12v motor for more than 2 seconds no matter what happens. If something goes wrong your hardware will be preserved for use in tele-op. Of course the amount of time you wait in your specific code depends on how long it normally takes the actuator to move into position. You should give it more than its normal amount of time to move into position, but keep the maximum time below the approximately 5 second stall time that can be expected to start damaging the motor. For example, after running several tests of the actuator you find it normally moves its full extent in 1.5 seconds. You might in that case allow the timer to run for 3 seconds before cutting it off for safety. This is below the normal amount of time that would damage the motor by overheating, but gives the actuator plenty of time to make it into position even if there is a minor problem.

 # 244. TRICK: Encoders tend to be fragile and don't work well in the face of slipping, so are not as useful as you might think for measuring motion.

We have not had good luck using motor encoders for control. We have found that they tend to break when a collision puts stress on the motor shaft. The usual purpose of an encoder is to measure the angle through which the motor has turned, for example ensuring the robot has gone forward a certain distance in autonomous mode. Unfortunately, because of wheel slippage, the encoder may not be fully accurate unless the motors are run at relatively slow speeds.

 # 245. SECRET: If you experience frequent NXT brick crashes and you are using encoders, try removing the encoders.

In the 2010/11 season there was some evidence that use of encoders could increase the frequency of NXT brick "crashes." The kind of crash we are talking about here involves the NXT brick displaying the last screen full of information it received from the Field Management System, and locking out all buttons. The only way to reboot the brick in this kind of crash is to physically remove the battery.

Although this has not been definitively proven to be an encoder issue, we did notice that our robots that used encoders seemed to crash more frequently than those that did not. The theory is that changes in the I2C driver code in RobotC in 2010/11 may have increased the frequency of NXT crashes and encoders apparently cause a great deal of I2C traffic. But if you have a lot of NXT crash events and you are using encoders, we would recommend trying the robot with encoders disconnected to see if that improves the situation.

Obviously whatever mechanism you were using the encoders to control would need to be reworked somehow, for example using limit switches for an actuator, using dead reckoning or gyroscopes for autonomous drive control, etc.

Note: As of this writing, the problem has been fixed but whatever the fix was has not been announced.

#246. SECRET: When using encoders to position heavy loads sometimes nMotorEncoderTarget won't ever get there; use custom code instead.

The nMotorEncoderTarget function reduces the amount of power the closer the motor gets to the target reading. If the load you are moving is heavy, there may actually not be enough power to quite reach the target. In that case, if you were waiting in a loop for the encoder to reach the target value, this may cause you to wait forever, since it will never get there!

If you find this is a problem, try using this custom code, which allows you to adjust the amount the power changes based on the distance from the target.

```
void MoveToPos(int npos, tMotor nmport, int range)
{
bFloatDuringInactiveMotorPWM=false; // prevents drifting
nMotorEncoder[nmport]=0;  // start the encoder at 0

// while the motor isn't within the acceptable range
  while((nMotorEncoder[nmport]<(npos+range))||
nMotorEncoder[nmport]>(npos-range))) {

    // adjust the power to the motor to be proportional
    // to the distance from the target

motor[nmport]=(nMotorEncoder[nmport]-npos)* 5;

    // NOTE: If this moves the actuator in the wrong direction
    // then change the - to a +  in the code above.
    // If it moves too fast make the 5 smaller
```

```
    // or if it moves too slow or never gets there then
    // make the 5 larger.
}

motor[nmport]=0;
}
```

 # 247. TIP: Test light sensors under different lighting conditions.

If you are using light sensors, for example to follow a line, be sure you try it out under different lighting conditions. You probably will need to "calibrate" your light sensor when you arrive at a tournament, since the field there may have different lighting levels than the one you've practiced on.

 # 248. TRICK: The gyroscope offset is usually closer to 600 than the documented 620; however you should calibrate the gyroscope in your initialize function.

The gyroscope documentation says it has a "zero" value of about 620, but in our tests we find it is usually closer to 600. In any case, calibration of the gyroscope zero value is highly recommended. This is most conveniently done in the intializeRobot() function. Note that the robot must not be moving while the gyroscope calibrates. To calibrate the gyroscope, you simply take the average of several readings made ten or twenty milliseconds apart from each other. You then store this averaged value in a global variable and subtract it from raw analog readings. This finds the "zero" point of the gyro, which can change based on temperature, "drift," and other factors.

 # 249. SECRET: The gyroscope is the most accurate way we've found to make turns.

The Hightechnic Gyroscope sensor is a marvelous device. It can measure "rotational velocity" in degrees per second. That might not sound too useful, but with proper programming you can "integrate" its readings and get surprisingly accurate information on how far your robot has turned. In our tests it is normally accurate within a single degree. And unlike the motor encoder, it is directly reading the rotation of the robot, so wheels slipping on the floor do not affect it in the slightest. In fact, during one practice round one of our wheel hubs came loose. Even though one drive wheel was free-wheeling with no power at all, the gyro compensated perfectly (the turn was very slow!) and the robot still scored 4 out of 8 balls in the high goal.

The following code shows how to integrate values to make a turn. The GyroRead() function returns the current value of the gyroscope (the way you read the gyro differs depending on whether you are using a sensor multiplexor or not).

Note that the speed of the turn, exact type of drive train, how you have mounted the gyroscope, and other factors greatly affects accuracy and you should run your own accuracy tests before deciding to use this technique.

```
// integrate readings from the gyro and return when the
// gyro hasturned the number of degreesspecified on the
// input parameter "angle". Note that if the gyro isnot
// located exactly in the center of the robot's wheels
// this angle won't correspond to the robot's actual angle
// of turn, but it should be consistent and repeatable from
// one run to the next. Positive angle is clockwise,
// negative counter clockwise.

int
GyroTurn(int angle){
 long curAngle = 0;
 int gyro = 0;
 while(true){
   // because we are reading every 50 milliseconds
   // that's 20readings per second, therefore each reading
   // represents a twentieth of a degree per second.
   // So multiply the angle by 20 to ensure we are
   // measuring using the same basis

   if (angle>0 && curAngle>angle*20 )
          break;
   if (angle<0 && curAngle<angle*20)
          break;
   curAngle += GyroRead();   // GyroRead() is a function you write
                             // that returns the current gyro val
   wait1Msec(50);   // if you change this to a number other
                    // than 50milliseconds you also must
                    // multiply the angle by the reciprocal of
                    // the time delay in seconds. For
                    // example if this was 100 milliseconds
                    // you'd multiply angle by 10 above instead
                    // of 20
 }
 return 0;
```

 # 250. TIP: Ultrasonic range finders work very well as long as you are pointing directly at a relatively flat, hard object.

We have found the Ultrasonic Rangefinder to be accurate within a centimeter or two in most cases, as long as it is pointed directly at a relatively flat, hard surface. If the surface is angled with respect to the robot, the sound waves tend to bounce away from the detector and accuracy is compromised; in some cases the object may not be detected at all.

 # 251. TIP: Ultrasonic range finders will read -1 if you are too close.

Be aware of this limitation! If you are very close to an object (less than a few inches) the range finder may fail and return -1 as the distance. You can avoid this by mounting it somewhat within the body of your robot, making it impossible for it to ever get too close. In any case, your program should take into account this possible reading and take appropriate action if -1 is returned from the sensor.

 # 252. SECRET: Use single "ping" code to reduce ultrasonic range finder interference if you are using more than one.

A disadvantage of ultrasonic range finders is that if you are using more than one of them then they can sometimes get "confused" by interference from one another. In other words, a "ping" coming from one will bounce back and be detected by the other one, causing it to sense an incorrect distance.

The code below causes each ultrasonic range finder to send out a single ping then wait for it and return a single value, rather than sending out a continuous stream of pings. This means one range finder will finish reading its ping before the other one starts, eliminating interference.

```
// Call InitializeUSSensor once for each ultrasonic range finder
void InitalizeUSSensor(tsensors nport)
{
static const byte kSonarInitialize[]={3, 0x02, 0x41, 0x01};
SensorType[nport]=sensorI2CCustomStd9V;
SendI2CMsg(nport, kSonarInitialize[0],0);
wait10msec(5);
}

// Call PingUSSensor each time you want a reading. It does not
// return until the ping is read or times out.
void PingUSSensor(tsensors nport)
{
// kSonarPing[]: data to tell the sensor to ping once
static const byte kSonarPing[]={3, 0x02, 0x41, 0x01};
```

```
// kSonarRead" data to tell the sensor to report its readings
static const byte kSonarRead[]={2, 0x02, 0x42};
const int nSonarReplysize =1;

byte replymsg [1];

SendI2CMsg(nport, kSonarPing[0], 0);
wait1msec(10);

nI2CBytesReady[nport]=0;  // Clears any pending bytes
SendI2CMsg(nport, kSonarRead[0], nSonarReplysize);
while (nI2CStatus[nport]==STAT_COMM_PENDING){
wait1msec(2);  // Wait till comm ends
}
readI2CReply(nport, replymsg[0], nSonarReplysize); // Read reply
 SensorValue[nport]=replymsg[0];  // Store the value
}

// Below is an example of how to set up and get values from
// two ultrasonic sensors using the one-ping-at-a-time functions.
// This could also be used for more than two.

const tsensor kUS1=S1;  // these could also be set up using
const tsensor kUS2=S2;  // pragma statements in the config wizard

task main(){

  int US1Val, US2Val // Store UltraSonic values
InitializeUSSensor(kUS1);
InitializeUSSensor(kUS2);


  // later in the code to get readings from two sensors do this:

PingUSSensor(kUS1);
PingUSSensor(kUS2);
US1Val=SensorValue[kUS1];
US2Val=SensorValue[kUS2];
```

 # 253. TRICK: IR Seeker Sensors require special setup in RobotC.

In order to use the IR Seeker sensor, you must have it set up in RobotC. Using the Motors and Sensors Setup screen, add a HiTechnicIRSeeker1200 sensor on the port you want. (Note that you must check the "Allow 3rd Party Sensors from HiTechnic" box to add the IR sensor.) It will add a pragma line like this:

```
#pragma config(Sensor, S1, IR, sensorHiTechnicIRSeeker1200)
```

You can then get the number of the zone that the IR beacon is in with:

```
int value = SensorValue[IR];
```

Where IR is the name of the sensor. (For more information on the IR Seeker's zones, consult HiTechnic's documentation.) For example, the following program will print out the IR Seeker's zone number on line 2 of the NXT's screen, where the IR Seeker is on port 1 of the brick:

```
#pragma config(Sensor, S1, IR, sensorHiTechnicIRSeeker1200)

task main()

{
  nxtDisplayCenteredTextLine(2,"IR Sensor Value:"+SensorValue[IR]);
}
```

 # 254. TRICK: Be careful when using the Sensor Multiplexor (SMUX); understand its limitations.

If you have more than three sensors you will need to use a Prototyping Board, a Sensor Multiplexor (SMUX), or both. This is because the NXT Brick has only 4 sensor ports and you must use one of those sensor ports for your motor controllers (you can daisy-chain all of the motor and servo controllers on a single port in most cases). This leaves only three ports for sensors. Sophisticated robots frequently use more than three sensors.

But if you do use the SMUX, be prepared for a few limitations. First of all, we have found it does not always work reliably for sensors that require two-way communications. For example, don't put a Prototyping Board on the SMUX, give the Prototyping Board its own sensor port. The SMUX can sometimes use standard code for accessing sensor readings, but sometimes it cannot and you must instead create your own custom code for reading the sensor values. For example, analog sensors such as a light sensor can only be used with the SMUX by reading the raw analog value and then scaling it: you cannot use the usual light sensor functions. We also found that the gyro sensor could not be used reliably with the standard gyro functions when it was placed on a SMUX, instead we had to write our own calibration function and a function to scale the output. The SMUX also explicitly does not support every sensor, see the description of the Sensor Multiplexor on the HiTechnic website for a list of supported sensors.

Future versions of the SMUX and supporting SMUX functions may correct these limitations. But be forewarned that if you plan to use the SMUX start using it early in the season so you can learn about and compensate for its limitations.

All of the forgoing is not intended to discourage you from using the SMUX. We use it and it is essential if you are going to use more than just a few sensors. Just be aware that you won't just plug your sensors in the SMUX and have them work just like they did before, it takes more work than that.

 # 255. TIP: Put a label on the SMUX battery case to make it easier to tell which way to flip the switch to turn it on and off.

The SMUX has its own battery case that takes a 9v battery. The switch on this case is quite difficult to read in dark conditions. We place a white stick-on label next to the switch to indicate which direction is ON vs. OFF. This makes it much easier to visually confirm the SMUX is turned on. Failing to turn it on results in all of your multiplexed sensors becoming inoperable, so be sure to check it.

 # 256. TIP: Be careful about turning the SMUX battery on and off.

Failing to turn the SMUX off at the end of the day will lead to the frustration of frequently dead batteries. Failing to turn it on before a match will mean all your multiplexed sensors will be dead. Make the SMUX battery a part of your checklist. Before each tournament check the battery level using a multimeter and replace it if it's weak. We have found the SMUX battery lasts quite a long time as long as you're using mostly passive sensors, but you should make sure that you have a spare 9v battery with you at any tournament if you're using the SMUX.

 # 257. TIP: There is a difference between the Sensor Multiplexor (SMUX) and Touch Multiplexor.

The Touch Multiplexor can only accept touch sensors and nothing else. The Sensor Multiplexor can accept many different types of sensor, including other Sensor Multiplexors, but there are some limits on what it is able to handle so be sure to read the documentation carefully.

Using the 3rd Party Sensor drivers

The 3rd party sensor drivers are a collection of header files that allow you to use the advanced functionality of many digital sensors, including many made by Hitechnic, Mindsensors, and Lego.

 # 258. HINT: Enable RobotC 3rd party sensor drivers.

Note: For this tip, you must be on the Expert or Super User Menu levels. To do this, go to the menu View/Preferences/Detailed Preferences. Open the Directories tab. In the box labeled "Directory for include" type:

```
.\Sample Programs\NXT\3rd Party Sensor Drivers\
```

(The last slash is important!) Click the OK button.

This will allow you to use the 3rd Party Sensor Drivers that are included with RobotC.

 # 259. HINT: When using the 3rd Party Sensor Drivers, be sure to choose the right sensor type

When you use the 3rd Party Sensor Drivers, set the sensor to be one of the following times:

- sensorI2CCustom
- sensorI2CCustom9V
- sensorI2CCustomFast
- sensorI2CCustomFast9V
- sensorI2CCustomFastSkipStates9V
- sensorI2CCustomFastSkipStates.

In the motors and sensors setup screen, check the "Allow Custom Designed Sensors" box under the Device Mgmt NXT tab (Requires the Expert or Super User menu level), and select one of the sensor types beginning with I2C Custom.

Or, If you're writing your own pragmas, write a line like this:

```
#pragma config(Sensor, S1, colorsensor, sensorI2CCustom)
```

You can replace sensorI2CCustom with one of the other sensor types listed above, if you like.

Each of the types of I2C sensors mentioned above is different.

Sensor type name in the Motor and Sensor Setup Screen	**Sensor type name in the pragma**	**Description**	**Readings per second**
I2C Custom	sensorI2CCustom	standard I2C	111
I2C Custom Fast	sensorI2CCustomFast	A sensor type with an faster baud rate	249
I2C Custom Faster	sensorI2CCustomFastSkipStates	Same as sensorI2CCustomFast, but it removes some logic required to read certain sensors. Fastest of the three.	333

The readings per second value is the maximum number of readings each second on a Hitechnic Color Sensor V1. This value was achieved by experimentation, and your results may vary.

In general, we recommend using sensorI2CCustomFastSkipStates on any digital sensor (when using 3rd Party Drivers) because of the greatly reduced amount of time it takes to get a sensor reading. Sensors often require very fast readings for time sensitive work. Certain sensors don't work on this setting, so if you experience problems, try using the sensorI2CCustomFast or sensorI2CCustom types instead.

Each of the above sensor types also has a 9v version for use with the Prototype board.

As an example, we'll look into reading values for the Hitechnic Color Sensor.

Without using the 3rd Party Drivers:

```
#pragma config(Sensor, S1, colorsensor, sensorI2CHiTechnicColor)

task main(){
    while(1){
        nxtDisplayCenteredTextLine(0, ""+SensorValue[colorsensor]);
    }
}
```

This is a basic example that simply outputs the Color's ID number to the screen. Consult Hitechnic's documentation for more information on which ID corresponds to what color.

It uses RobotC's built in drivers to read data from the sensor. Note that when using the internal drivers, you set the sensor type to sensorI2CHiTechnicColor (or whatever type corresponds to the type of sensor you have), and then simply read values from SensorValue[].

Using the 3rd party drivers:

```
#pragma config(Sensor,S1,colorsensor,sensorI2CCustomFastSkipStates)

#include "drivers/HTCS-driver.h"

task main(){
    while(1){
nxtDisplayCenteredTextLine(0,
""+HTCSreadColor(colorsensor));
    }
}
```

This is the same as the previous example, except we're now using the 3rd Party Drivers instead of the RobotC Built-ins.

You can see that we have to set the sensor type to one of the Custom I2C types in the pragma, as mentioned in one of the previous tips. In order to actually use the drivers, you must include them. (HTCS stands for HiTechnic Color Sensor.) We now read the sensor's data from the driver's function, HTCSreadColor. The name of the driver file and the function name were determined by reading the documentation, available at http://rdpartyrobotcdr.sourceforge.net/

In our third example, we start using some of the additional functionality of the 3rd Party Drivers:

```
#pragma config(Sensor,S1,colorsensor,sensorI2CCustomFastSkipStates)
```

```
#include "drivers/HTCS-driver.h"

task main(){
  int r, g, b;
  while (1){
    HTCSreadRGB(colorsensor, r, g, b);

    nxtDisplayCenteredTextLine(0, ""+r);
    nxtDisplayCenteredTextLine(1, ""+g);
    nxtDisplayCenteredTextLine(2, ""+b);
  }
}
```

Rather than reading a color index, this reads the RGB components of the color the light sensor currently sees. The variables r, g, and b are passed by reference, not by value, to the HTCSreadRGB() function (You can tell because of the & in front of the variable in the documentation). The function modifies the variables, and when it returns, the variables each have the value of that given color component. We then print them to the screen.

Dead Reckoning

Dead reckoning is a technique in which your robot makes moves entirely by turning on the motors for a set period of time. In other words, you turn the motors on at a certain power level, you wait some number of milliseconds, and then you turn them off. To go straight you move them all in the same direction; to turn you move one side's motors in the opposite direction. You are not using any sensors to help guide the robot, you are simply trusting that repeating the same motions gets you approximately where you need to go.

This is the simplest way to move your robot from one place to another during autonomous mode. If the number of motions you need to execute is small, or if accuracy is not critical (for example, a defensive autonomous mode that simply tries park somewhere to interfere with an opponent's autonomous mode only needs to be accurate to within a few inches), then dead reckoning can be good enough to perform the task successfully. And this technique certainly observes the tip that "simplicity is good."

 # 260. SECRET: Dead reckoning may be accurate enough if you have only a single turn to make.

We frequently have at least part of our autonomous use simple "dead reckoning," especially for forward motion. We have found that turns can be a problem using dead reckoning and prefer to use the gyro sensor for turn accuracy. However, if you are only making a single turn and it is not too severe (less than about 100 degrees) then even that can work out pretty well most of the time using only dead reckoning.

 # 261. TRICK: Measure the repeatability of your dead reckoning motions by marking wheel position with bits of tape.

It is important to understand how much variation there is when the robot moves using dead reckoning. Once you think you have the basic movements working, try a test run and mark where the four wheels touch the field mat using tiny pieces of tape. Then, run the same program several more times and keep track of how far the wheels are (on average) from those marks on each run. The smaller the distance, the better, of course. Try different techniques to reduce this variation, for example move the motors are different speeds, add small "settling pauses" after motions are complete, etc.

 # 262. STRATEGY: First get autonomous working as well as possible using only dead reckoning, then add sensors to increase accuracy if needed.

It is relatively easy to get dead reckoning working by sketching out the basic motions that get the robot to the target area then just adjusting time delays until it ends up approximately in the right place. In some cases a high score can be achieved just using dead reckoning alone. After you have done as well as you think you can using dead reckoning, do five or six runs and track the results. This is your baseline figure. Now, try adding sensors to help get to the target more accurately (for example by line following). Once that works, run your series of tests again to measure the performance using sensors. You may find that sensors add seemingly very little extra scoring power, or you may find the difference is dramatic. It is even possible that the use of sensors actually decreases the performance of autonomous by adding new time delays. The only way to know is to do careful testing.

 # 263. TIP: Even if dead reckoning works perfectly on your test field, variations in the tournament field may mean sensors are still desirable.

Let's say you've followed the TIP above and find you can score almost perfectly using dead reckoning alone. Should you just call it a day? Maybe not. Even if dead reckoning works very well on your practice field, there can be significant field variations at tournaments and scrimmages. In the 2009/10 season one of our robots consistently scored 6 to 7 out of 8 balls using only dead reckoning. But we found that at real tournaments the performance was much better using sensors. For example, in one round at the Pennsylvania Championship one of our opponents rammed our robot during autonomous, but the sensors compensated for the ramming and we still scored 5 out of 8 balls! That would not have happened using only dead reckoning.

Because of these effects, even if you see no improvement at all between dead reckoning and sensor based autonomous modes on your own test field, the sensor version of your program may perform better at actual competitions by compensating for field variations or defensive actions by your opponents.

Servo Programming

 # 264. TIP: You can't read the position of a servo. Some older RobotC documentation implies you can; it's wrong.

Once you set a servo to a position, you can't read the value again. For example:

```
servoTarget[myservo] = 42;
nxtDisplayCenteredBigTextLine(0, "" + servoTarget[myservo]); // Bad
```

You might expect this code to display 42 on the screen, but it doesn't. In the same way, the following code doesn't work:

```
servoTarget[myservo]++;   // Bad
```

Instead, the array servoValue[] holds the last target you set. So the following would work as expected:

```
servoTarget[myservo] = servoValue[myservo] + 1;
```

But in any case, note well that servoValue[] only repeats back to you the last target you specified for the servo, it says nothing about whether the servo has actually made it to that position yet. The servos in the FTC system are open-loop, meaning there is no feedback about what their actual position is. You would need to implement a limit switch or use a rotation sensor or encoder to confirm the actual position if that is necessary for your use of the servo.

 # 265. TIP: Every servo should have a reasonable value initialized in both autonomous and tele-op programs.

You don't want your program starting off servo motors in default positions. Determine the correct position of the servo and program it in to your robot initialize function. Share that function between both tele-op and autonomous using a #include file so they do not get out of sync with each other.

 # 266. TIP: Remember that the FTC servos are "clockwise" servos, meaning the position increments as the servo rotates clockwise assuming you are looking at the center screw of the servo horn.

Frequently you will hear programmers say things like, "I have that gripper servo set at 45 but that's too high; should I increase or decrease the value to bring it lower?" The answer is: if turning it clockwise brings the gripper mechanically lower, you would need to increment the value (say, from 45 to 50) but if you want it to turn counter-clockwise to achieve the desired result you would need to decrement it (for example from 45 to 40).

Simply remember it this way: The numbers pointed to by a clock's hands increase as it turns clockwise—so do a servo motor's servoTarget[] values!

 # 267. TIP: Use the Servo Debugger window in RobotC to calibrate servo positions.

The servo debugger window is a great tool in RobotC that allows you to test out various servo position values before actually writing your program. We strongly recommend using it. By finding and properly setting initial values on servo motors in the robot initialize function of your program, you will prevent servos from overheating; incorrect settings can cause servos to constantly consume power trying to reach a position that is mechanically impossible.

 # 268. TIP: Continuous rotation servos stop at 128, go clockwise at faster and faster speeds between 129 and 255, and go counter-clockwise at faster and faster speeds between 127 and 0.

The continuous rotation servos were introduced in the 2010/11 season and they can in effect be used as small motors. They are somewhat confusing to control because setting their target values controls speed and direction rather than absolute position as with standard servos. The mid-value (128) actually stops them from moving.

Higher values cause higher and higher clockwise speeds (when looking down at the center screw of the servo horn), while lower values cause faster and faster counter-clockwise rotation. A target of 0 actually goes as fast as possible counter-clockwise, which is also confusing for many people.

As with standard servos, the continuous rotation servos are open-loop, meaning there is no feedback about what their position is. You would need to use an encoder, a Hitechnic rotation sensor, a limit switch, or some other method to determine when they had reached a certain position.

Tele-Op Programming

Game Controller Layout

 # 269. TIP: Create a diagram showing controls to keep your drivers informed, and to train new drivers.

See the example in the picture. You can find pictures of game controllers online and either use a paint program to label them or just write it by hand. Print two copies so you can put one in your engineering notebook to explain the controls to judges.

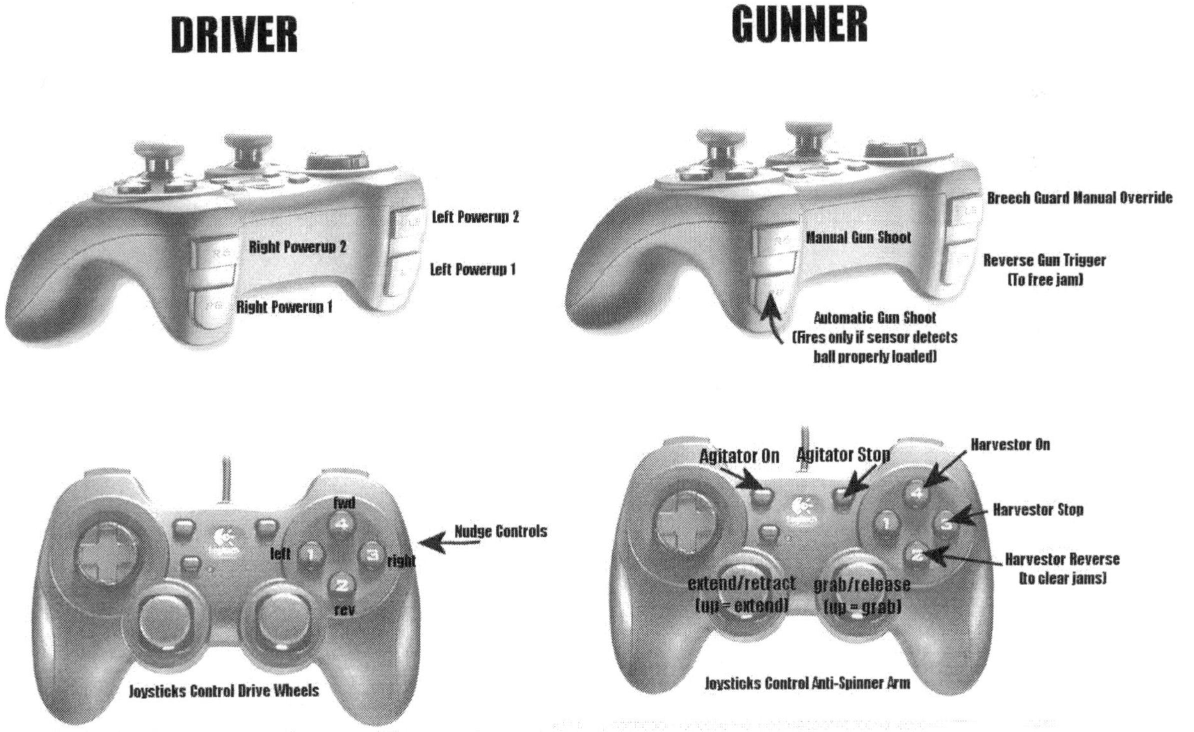

 # 270. TIP: Make the controls for different actions correspond to reality and logic as much as possible; make the controls "ergonomic."

For example, choose an orientation for the robot that is considered "forward." Perhaps there is a shooter pointing along the same direction as the wheels and your team will consider forward to be in the direction the shooter is facing. Now, make sure your joysticks controlling the drive train and all other controls are consistent with that orientation. Pushing the joysticks up is generally considered "forward" since the driver in effect imagines they are riding on the robot and using tank-like controls.

As another example, suppose you have an actuator that goes up and down. If you were going to use two buttons to control it, choose two buttons that are oriented one above the other, and make the one on top "up" and the one on bottom "down."

This may seem obvious but we've seen cases where programmers just arbitrarily selected buttons or directions that made little logical sense, and that simply increases training time for the drivers. In one case a programmer told the driver to "just get used to it this way," but that's not the right attitude! Take a few minutes and make it logical, you'll save everyone time and effort later. We have seen cases where a match was lost simply because a single wrong button was hit on the controller at a really bad time. Make things logical and that will happen less often.

 # 271. STRATEGY: Consider how many fingers the driver has and which functions must be activated at the same time when designing controls.

Think about coordination: if your control layout would require a driver or gunner to press seven buttons at once in a common situation that might be nearly impossible to achieve. As another example, having one joystick shoot and another drive the wheels might be hard to learn. Think of how video games use controls in ergonomic and easy to learn ways that may already be familiar to your driver team.

 # 272. TIP: Program "power up" buttons to allow drivers to select different speed levels.

The joysticks do not really give your drivers enough range to select slower speeds for fine maneuvers versus faster speeds to race across the floor. Typically in the heat of competition drivers will just end up "pegging" the joysticks all the way up or down. We have found it useful to have a scaling factor applied to the joystick inputs, using the front buttons (buttons 5, 6, 7, and 8) to select higher power levels. For example, with all of the front buttons unpressed, run the motors at 50% of the value of the joysticks. With the top front buttons pressed, run at 75% power, and with both pressed run at 100% power. You can allow these "power up" buttons to be independent on the left and right wheels if desired and if you don't need any of the front buttons for other purposes.

 # 273. TIP: Program "nudge controls" for fine-tuning.

In the 2009/10 season the game required the robots to shoot wiffle balls through the air at targets 30 inches off the ground. Positioning the robot for a shot was tricky—if you missed the target by an inch or two it was hard using standard controls to just slightly move to correct the aim. We programmed in "nudge" controls using buttons 1 through 4 on the controller. These controls would very slightly move the robot forward, backward, rotate left, or rotate right. This allowed for tiny little adjusting moves to be made easily. Many games could benefit from giving the driver a special set of fine controls for specific tasks like scoring.

 # 274. SECRET: Put the driving control code in a separate task so that the driver maintains movement control when other subroutines are running.

By putting the driving loop in a separate task, you don't have to worry about other subroutines that consume significant amounts of time locking your driver controls out.

To see how this could be a problem, take the example of a robot that has an actuator that requires 1 second to reach a certain position, and once it reaches that position you want to open a gripper, all as a single control. Your code to implement this action during teleop might target the main actuator servo to go to a certain position, wait 1 second, and then target the gripper to open. Let's say you place that action on button 4.

Now, let's say the driver is executing a turn and the gunner presses button 4. What's going to happen is that the driver's controls will effectively be locked out for a full second until the actuator function completes. The driver may be very surprised to find that he or she cannot stop turning during that one second. A second may not seem like a lot, but it can be enough to cause serious problems in a competition; for example causing the driver to crash into an obstacle or run off a ramp and tip over!

By placing the main driver control loop in a task, it will continue to run even as other functions are executed.

Now, you could also do the same thing by putting each and every time consuming action in its own task. But that causes a lot more programming headaches than simply putting the main driver loop in a task. It also may not be a good idea to have the NXT brick running too many tasks at once if you have many such actions.

The code below shows an example of how to do this, but obviously you may have very different sets of controls on your robot.

```
// Put the main driver control loop in its own task so
// the driver never loses control of the robot!

task drive()
{
  while (true){
    getJoystickSettings(joystick);

    if (abs(joystick.joy1_y1)>10) {   // dead zone
      motor[left]=joystick.joy1_y1;
    }else {
      motor[left]=0;
    }
    if (abs(joystick.joy1_y2)>10) { // dead zone
      motor[right]=joystick.joy1_y2;
    }else{
      motor[right]=0;
    }
  }
}

task main()
{
  initializeRobot();

  waitForStart();

  StartTask(drive);    // give driver control of the wheels

  while (true) {

     // Put code to handle all your other tele-op controls
     // in here, such as grippers, scoring systems, harvesters
     // etc. They can use wait commands for timing without
     // affecting the driver's ability to control the robot's
```

```
        // drive train.
    }

}
```

 # 275. TIP: Consider other types of automated functions during tele-op mode.

The joystick controls are not terribly accurate even after calibrating. Some maneuvers might be better implemented as pre-programmed automatic actions. For example, in the 2008/09 season, one of the two partner robots on a team had to come down off a steep ramp when the game started. We found it was quite easy to accidentally run off the side of the ramp by not precisely hitting both joysticks simultaneously. The reason was that you needed to run the robot at less than full speed, but joysticks are notoriously hard to run at exactly the same speed if less than maximum. Even our best drivers occasionally would flip the robot coming off the ramp. We programmed an automatic "deramp" button which would precisely run the motors at the same speed, and at a perfect speed for de-ramping. The automatic program was far more accurate than human drivers using inaccurate joysticks.

There may be similar operations in future games that are better solved using a special purpose tele-op function rather than relying on driver training using manual controls.

Buttons and Joysticks

 # 276. Call getjoysticksettings on every tele-op loop

There's often a noticeable lag between the driver controls and the robot in tele-op. This lag can be reduced, if you call:

```
getJoystickSettings(joystick);
```

at the beginning of the while loop in tele-op. While the controls will function even if you do not do this (because a background task calls this function every now and then) you will get much better response time if you call it yourself each time through the main control loop.

 # 277. TIP: Program a "dead zone" for joysticks to avoid motor "creep."

The joysticks are often not perfectly calibrated. This means that even when the driver is not touching the joystick, it may not read exactly zero. If you are using the joystick to directly select the motor speed, even if scaled, you might find that even when the driver is not touching the joysticks the robot will slowly move in one direction or another.

The solution is simple: after reading the joystick value, just consider anything within the range +10 to -10 to be zero. This only takes a few lines of code if you use the handy "absolute value" function abs():

```
// assume the joystick values have already been read in
// to the variables joy1value and joy2value

if ( abs(joy1value) < 10) {
      joy1value = 0;       // consider anything less than 10 to be 0
}
if ( abs(joy2value) < 10) {
      joy2value = 0;       // ditto
}
```

Support Functions

 # 278. TIP: Keep common functions that are used both by tele-op and autonomous programs stored in a file that is included by both.

This avoids duplication of code and when you need to adjust or change that common function; the improvement will automatically get picked up by both programs.

Note that in most cases such a file containing functions may give you warnings when you open it in the RobotC IDE since it may not contain pragma statements referenced within the functions. Just ignore this, the file does not need to compile on its own, it only needs to compile when it is included in the autonomous or tele-op programs.

 # 279. TRICK: Keep your motor and sensor pragma statements in a file that is included by both tele-op and autonomous programs.

By doing this you will avoid the common problem of hardware additions causing them to get out of sync. For example, if you add a new servo motor to your configuration, you might remember to update the pragmas in tele-op to reflect the new hardware configuration, but you might forget to update the pragmas in your autonomous program, which could cause errors. Especially subtle problems can occur if you change which channel an existing servo is plugged into (or port for a 12v motor, or sensor, etc.)

 # 280. TIP: Create common functions for motor control operations.

For example, MoveForward(50) might move the robot forward at 50% power. Turn(LEFT, 30) might turn the robot left at 30% power (where LEFT is a #define value). This makes the code easier to read than directly controlling each motor's power.

Calibration and Diagnostic Tools Programming

 # 281. TIP: Write a single stand alone program that can be used to calibrate light sensors, test internal and external battery voltage, and output other critical sensor values to the NXT screen.

This is more convenient than having lots of little programs scattered around. There is quite a bit of room on the NXT display to show information from several different sensors that may need to be tested or calibrated. It's helpful to output the voltages of the batteries as well.

 # 282. TIP: Create a stand-alone program that you leave on the NXT brick that runs the motors in sequence for a quick sanity check.

Sometimes it takes a while to boot up the laptop and establish connections with the robot. There may be occasions when you just want to quickly find out if all your motors are working. A simple way to do this is to store a small stand alone program on the NXT brick that just briefly fires up the motors in sequence, call it "motor test" or something similar.

The idea is, you set the robot on a stand or box that keeps its wheels off the floor, and then you run the motor test program. The program pauses for a few seconds, plays a warning tone, then runs each motor one by one for a second or two at a time, first forward then backward, with a few seconds pause in between each motor. The program may also trigger important actuator motors for short periods of time. Of course you must stand back out of the way when running such a program if it's going to trigger actuators, and you should make sure you have easy access to the kill switch or NXT off button in case something goes wrong and you need to terminate the program.

Don't run any motor more than a second or two, because if the motor stalls due to a hardware problem you could damage it by powering it too long.

Part VII: The HiTechnic Prototyping Board

The HiTechnic Prototyping Board (ProtoBoard) has many uses and is well worth the extra effort to study. Many teams, maybe most teams, are intimidated by this very powerful addition to your robot's sensor arsenal. But by following a few simple steps, anyone can use it effectively.

 # 283. TIP: A great place to get electronic components and sensor components is www.digikey.com (Digikey).

They have the most amazing electronic parts search in the universe! You can find all kinds of switches, sensors, potentiometers, and other components for your custom ProtoBoard circuits here.

 # 284. TIP: For inexpensive basic components, a great place to shop is www.elexp.com (Electronix Express).

While they do not have the huge variety and nifty search capability of DigiKey, Electronix Express has very low cost items like potentiometers, capacitors, LEDs, switches and buttons, multicolor wire, soldering kits, etc. Also check out their resistor kits that give you a huge assortment of standard resistors at a low cost.

 # 285. TIP: Another great place for electronic components is www.jameco.com (Jameco Electronics).

Jameco has been around a long time and has many interesting parts and materials.

 # 286. TIP: Another great place for electronic components is www.goldmine-elec.com (The Electronic Goldmine).

This is an excellent site that sells groups of surplus electronics at low prices.

 # 287. TIP: Low cost PC boards with solder pads can be obtained inexpensively from www.radioshack.com (Radio Shack).

They have other components as well, but where Radio Shack beats the other sites is low cost drilled PC boards with solder pads. Search for "PC Board" to find many suitable items. You want one at least a few inches long by a few inches wide. Be careful to get a pad hole pattern that fits your project's needs. We prefer simple grid style patterns for ProtoBoard work.

 # 288. SECRET: Use the solderless breadboard version of the protoboard, not the direct solder version.

There are two versions of the HiTechnic ProtoBoard. One of them is built so you can directly solder components onto the board. The second one has pins that can insert into standard 0.1 inch spaced breadboard systems.

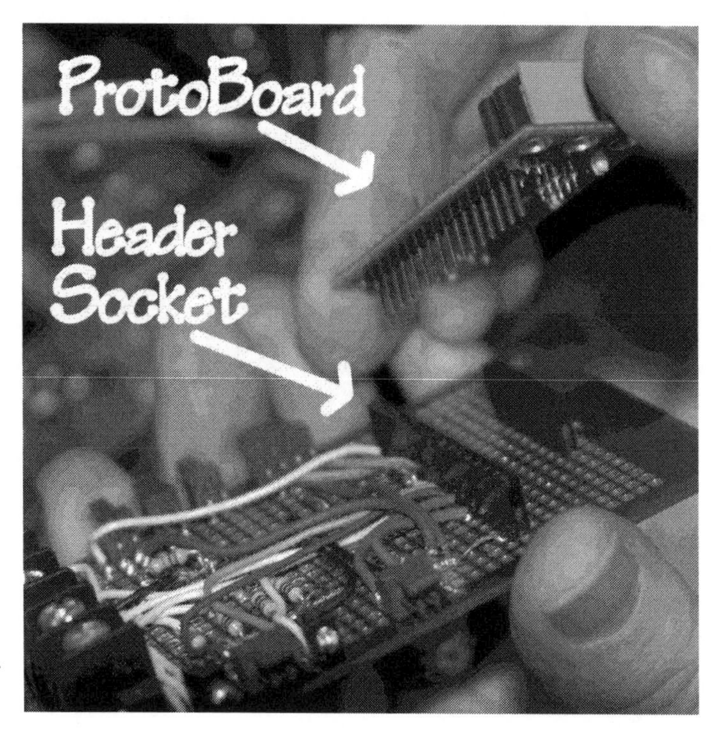

We recommend you always buy the solderless breadboard version. You can purchase 0.1 inch female headers that snugly accommodate the male pins on the Prototyping Board. In the picture you can see we also put an unconnected header behind the one meant to receive the ProtoBoard pins; that one is there simply to give the Prototyping Board something to rest on.

The reason for this tip is that if you accidentally static discharge onto your ProtoBoard and burn it out, you can easily replace it. The components you solder onto your circuit board are likely to be less static sensitive than the microchips on the ProtoBoard (typically you will use components like resistors, switches, terminal strips, and maybe LEDs, none of which are very sensitive to static electricity.) So, instead of having to rewire your entire project onto a new ProtoBoard that you can solder, you merely unplug the dead board and swap in a new one.

 # 289. TRICK: Completely cover the solder side of your board with electrical tape.

There are lots of opportunities to short out your board accidentally on the robot's metal structures. The rules allow you to use electrical tape to insulate any electrical connection, and this is a great way to protect your hard work on the Prototyping Board. Make sure no components puncture through the tape. We also created a plastic holder for our ProtoBoard inside the robot. The board was held sandwiched between two pieces of plastic and could be slipped in and out. This provided further protection against accidental shorts.

 # 290. TRICK: Ground yourself before touching the proto board.

The ProtoBoard is static sensitive. The material used to cover the field floor can build up significant static electricity. Try to always touch something large and metal before touching the ProtoBoard to discharge yourself.

 # 291. TRICK: Use low power LEDs on the soldered board to confirm switch closures.

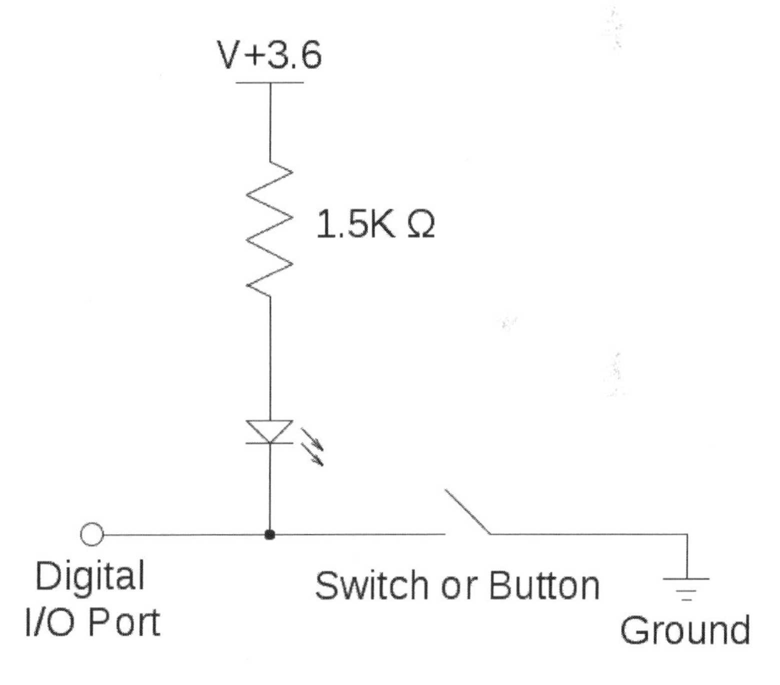

If your ProtoBoard is being used to detect switch closures of some kind (snap switches, push buttons, etc.) then it can be very useful to include a very small, low power LED right next to the incoming wires for the switch or button. This allows you to confirm the wires and switch are functioning simply by manually throwing the switch then watching to see if the LED lights. This can save loads of debugging time. In such an application, you will use a digital IO line and you will pull the line to ground when the switch is thrown. Be sure to include a suitable sized resistor so the LED does not draw too much power. We use tiny LEDs that light up with only two to three milliamps of current. Since the IO lines are on a 3.6 volt circuit, this means you need resistance of about 1K ohms to 2K ohms. The circuit diagram here shows one way of having a switch both register on a digital IO port as well as lighting an LED. Note that when the switch is closed the IO port will be grounded and will read zero, and when the switch is open the IO port will be pulled to +3.6 volts via a weak pull-up resistor and will read one. This is the opposite of what you normally think: open is one and closed is zero.

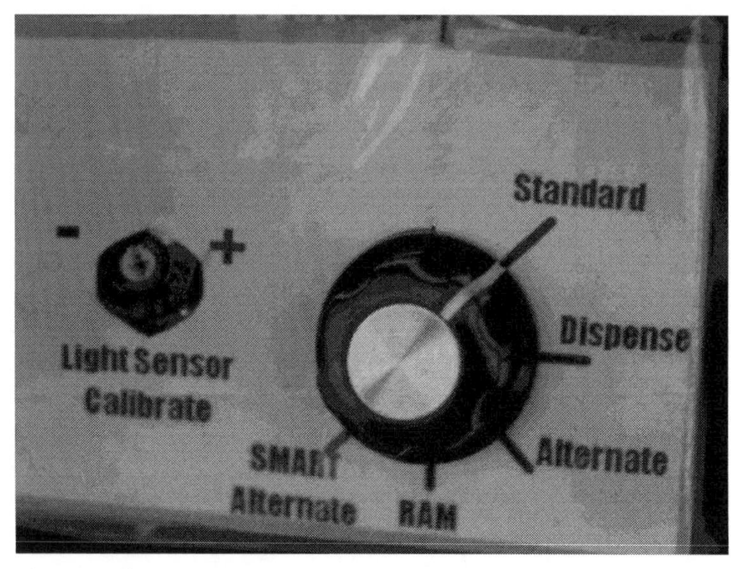

 # 292. TRICK: Use potentiometers on analog ports or slide switches on digital ports to select from many possible autonomous programs.

If you have many different modes for your autonomous program, it can be error prone and confusing to try selecting all the options using the NXT brick buttons or other methods. By using an analog port on the ProtoBoard along with a simple 20k to 40k potentiometer, you can easily select from six or seven different choices.

The picture here shows two potentiometers being used, one allows the operator to calibrate the light sensor without modifying the program's constants (see the following TIP for details), and the other allows one of five different major autonomous modes to be selected. (Standard tries scoring the starting balls in the high goal, Dispense tries to dispense first then score those balls in the high goal, Alternate tries scoring from a secondary field position that is less subject to rams by the opponent's autonomous, RAM tries to ram an opponent to stop their autonomous from working, SMART Alternate uses an ultrasonic sensor to detect an oncoming ram attempt by an opponent and automatically goes to the alternate field scoring position.)

The second picture shows the use of two slide switches. The one on the left selects left or right field position (and to reduce operator error it is positioned sidewise so the physical switch position corresponds to left and right), the one on the right selects whether the harvester runs during autonomous, up is on and down is off (sometimes it was advantageous to store additional balls in the harvester for use only

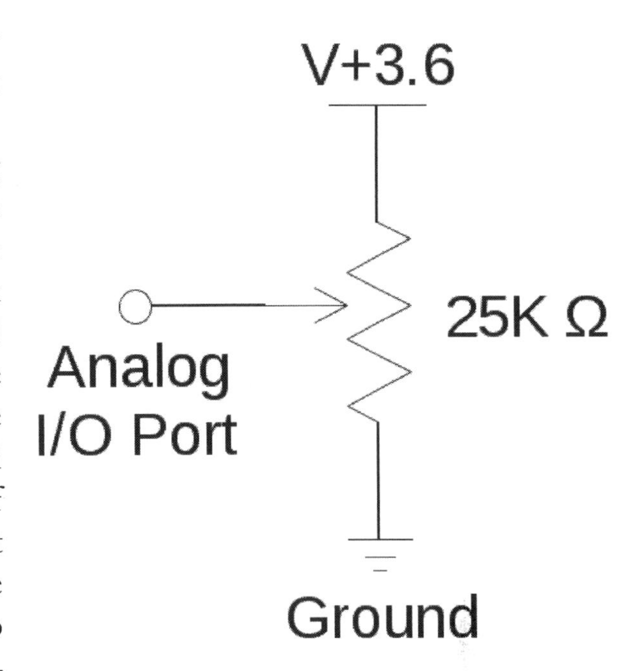

in tele-op, by not running the harvester they were retained for tele-op rather than scored in autonomous.)

The circuit diagram for interfacing with a potentiometer is shown in the diagram. A potentiometer (or "variable resistor") whose value is between 20k ohms and 50k ohms works well and does not draw very much current. The circuit is a simple voltage divider whose output voltage will vary between 0 and 3.6 volts depending on the position of the potentiometer's dial. The output comes from the central terminal of the potentiometer and may be connected to any of the analog ports on the Proto-Board.

The code to read the potentiometer and translate it into one of the desired values is very simple. You just read the analog port value then you divvy up the range 0 to 1024 into sub-ranges that correspond with the switch settings you desire. You consider any value within, say, 100 of the center of the range to be a particular value. For example, in our autonomous mode selector we use readings of 0 to 200 for "Standard," 201 to 400 for "Dispense," 401 to 600 for "Alternate," etc. A test program that simply dumps the analog value to the screen is used to calibrate the potentiometer, setting each selection to the center of the range.

We could have easily added two or even three more settings by using narrower ranges. However, if you try to select too many different settings from a single potentiometer you will eventually get into trouble because the ranges will be so narrow it will be easy to "drift" into the wrong setting. Remember that potentiometer readings are also affected by ambient temperature and even humidity to some extent, so dividing up the range into too tiny slices might result in inconsistent behavior depending on external conditions.

Here is example code:

```
// defined constants for different autonomous modes

#define MODE_STANDARD 0     // score high goal, near position
#define MODE_DISPENSE 1     // dispense first then score
#define MODE_ALTERNATE 2    // score high goal, far position
```

```
#define MODE_RAM 3          // Ram opponent's auton. position
#define MODE_SMART 4          // detect ram attempt, score close or
far

// Read the ProtoBoard ADC port number 3, 10 bits

int autonomousSelector = HTPBreadADC(HTPB,3,10);

// divide up the potentiometer reading into ranges that select
// different modes

if (autonomousSelector < 201) {
     mode = MODE_STANDARD;
} else if (autonomousSelector > 200 && autonomousSelector < 401) {
     mode = MODE_DISPENSE;
} else if (autonomousSelector > 400 && autonomousSelector < 601) {
     mode = MODE_ALTERNATE;
} else if (autonomousSelector > 600 && autonomousSelector < 801) {
     mode = MODE_RAM;
} else if (autonomousSelector > 800) {
     mode = MODE_SMART;
}

// Later in the program the "mode" variable is tested to
// select which actions are performed depending on the desired mode
```

 # 293. TIP: A potentiometer can also be used to calibrate a light sensor without having to modify your program.

As shown in the prior TIP, we also use a potentiometer to calibrate a light sensor. Normally we see teams calibrate light sensors by running a calibration program (preferably on the exact field you will be competing on) and then modifying a threshold constant in their program. They then must recompile the program and download it to the NXT brick. This is somewhat time consuming and error prone.

Instead, simply use another analog port on the ProtoBoard to get a value from a potentiometer, and use that value to modify your light sensor threshold. The same circuit as used in the prior TIP can be used for this purpose.

The simplest way we've found to use the value is to simply scale it to the range of 0 to 100 and use that directly as your light sensor calibration value. For example:

```
int lightThreshold = HTPBreadADC(HTPB,2,10); // ADC #2, 10 bits

if (lightThreshold == -1) {     // proto board failed!
     lightThreshold = 35;       // use a reasonable default
} else {
```

```
        // scale to range 0-100
        lightThreshold = (lightThreshold*100)/1024;
}
```

 # 294. SECRET: Building custom sensors can be cheaper and better performing than using standard sensors.

This is hard to believe but it is true! If you search through DigiKey, you will find all kinds of amazing sensors including the standard ones you can purchase from Lego and HiTechnic, such as ultrasonic range finders, snap switches that make great touch sensors, or even multi-axis gyroscopes. Frequently, you will find that buying the components and wiring up the sensor yourself using the ProtoBoard is far cheaper than the off the shelf sensors. Not only that, but you can pick the exact custom part you need.

For example, you can find many different types of ultrasonic rangefinder, often for half the price of the Lego version, but with different specifications (different detection ranges, higher accuracy, etc.) The same goes for gyroscopes. You can get dual axis gyroscope chips for under $10 that are more accurate than the one from HiTechnic and a fifth the price. Of course, you have work to do when you build your own sensor, and you need to construct some kind of housing for it, so this is less convenient than using the commercial sensors. But then again, you will learn a lot by building this kind of project, and learning is what FTC is all about!

One thing to be wary of is that you will need a very steady hand to wire up tiny surface mount chips that are commonly sold for many kinds of sensor. Where possible, try to get through-hole chips instead. There are videos on Youtube that illustrate how to solder the tiny surface mount chips, however, so it can be done.

It is simplest to interface with sensor chips that already have an analog output. Those that output using an I2C protocol will be a little more challenging on the software side, but they are still possible to use.

So try building your own sensors, you might be surprised at how well they work!

Part VIII: Training

As in any sport, you need to train and practice before even the first scrimmage. This gives you an edge over the competition, and allows your team members to respond to common situations with confidence. But it's not just your drivers who need to practice! As we'll see in the tips in this chapter, the entire team must practice ahead of your first competition.

Driver Training

#295. STRATEGY: Design a series of timed drills based on the game. Run these drills and track progress of each driver team over the course of the season.

In the same way that you test the robot by designing repeatable tests, you also need to be able to judge how well the drivers improve over time. Designing a set of tests based on the season's game gives the drivers a way to measure their own progress and push themselves to achieve better scores.

For example, in the 2009/10 season we had five different drills that were often used as warm ups by the drivers, and we tracked their scores. In one drill, the robot started with 8 balls in the hopper and the drivers had 1 minute to score as many as possible in the high goal. They were not allowed to harvest so accuracy was important, but they were also timed so speed was important too. Similar drills were used for the off field goal, and for harvesting, and even just racing around the field as quickly as possible to help them learn the controls. By tracking the results on a large pad of paper, we could see how the drivers improved as well as how hardware changes affected the driver's ability to perform critical game tasks.

#296. STRATEGY: Build a "sparring partner" robot.

If your school is like most and you have only a single FTC team, then you need to build at least a simple "sparring partner" robot. This robot can use minimal materials and doesn't even need to be able to harvest or score. Its purpose is to run around on the field and play defense against your drivers so they learn how to deal with opposing robots.

 # 297. STRATEGY: Hold joint practice sessions with other teams in your area.

Although you will reveal your robot's strategy to one potential opponent, generally speaking you are still better off getting real life practice playing the game against a real opponent before your first tournament. You will learn things that other teams who do not follow this tip will only figure out by potentially losing rounds that actually count!

 # 298. TRICK: Play the way you practice and practice the way you want to play.

The strategies and techniques you work on during practice should actually be the ones you really want to deploy in real tournament situations. That may seem obvious, but sometimes drivers lose confidence during real matches. For example, in the 2009/10 season there was a yellow "doubler ball" that would double the score of the goal it ended up in, but it could only be deployed and shot during the final 30 seconds of the match. Many teams practiced using the doubler ball extensively, but in the heat of actual competition decided not to use it for one reason or another. The point is, if they were not going to actually use it in a real tournament, why waste time practicing it? Why not practice other techniques and strategies that were actually going to be employed? Practice time is limited; it is a precious resource.

 # 299. TRICK: Design strategies for dealing with different kinds of hardware failure and frequently review these strategies with the driver team.

Thinking about what to do if different kinds of failure occur, and then reviewing possible ways of dealing with those failures is one way to make the best of a bad situation. When drivers are out competing against real opponents they are under a great deal of pressure. When hardware fails, the pressure only becomes more intense, and this is the time when drivers make mistakes. Talking about what to do ahead of time allows drivers to remain cool and make rapid decisions to try to recover.

For example, in the 2009/10 season a team could score 5 points by shooting a wiffle ball in a high goal, or could wait until the last 30 seconds and shoot for 10 points into off-field goals. Using the "multi-dimensional threat" tip, our strategy was to score both ways: score some high goals, harvest more balls, and then score those additional balls off-field at the end. That was the strategy if the hardware all worked as designed.

But what if hardware failed? We quizzed the drivers: "With 90 seconds to go you have 5 balls in the hopper and your harvester chain pops off. You can't harvest. What do you do?" The drivers thought about it for a little while then said, "Save the 5 balls for the end game since they can score 10 points there and you can't harvest any replacements. That maximizes your possible score." That seems obvious when you think about it for a few moments, but out on that field after the stress of a hardware failure, you have no time to think, you must make a split second decision. By reviewing possible situations ahead of time, drivers were always aware of what to do, and didn't have to think about it.

Of course, in some failure situations there isn't much you can really do to recover, but it's useful to review those anyway. "What would you do if you had 10 balls in your hopper, 90 seconds to go, and your shooter motor failed?" The answer, which took some time to figure out: "Play defense until the very end." A possible strategy of trying to purposely tip the robot over to spill balls into the low goal was rejected: the maneuver was too unlikely to produce many points; playing defense to prevent even a single opponent ball from scoring off field was superior to a desperate attempt that could only score a handful of points at best.

Pit Crew Training

 # 300. TIP: During build sessions when common problems occur that need to be fixed, time the pit crew to see how long the repair takes.

For example, a burned out motor is a common problem. Knowing how long it takes your team to repair that problem will be useful during tournaments when you may have limited time to make a repair.

 # 301. TIP: Make a checklist of items to review between rounds.

Making a checklist of things to do before each round is critical. Robotics teams, even those who have built the most advanced robots, often forget very small tasks like checking for loose screws or lubricating critical mechanisms. You'd hate to lose a round because a single screw fell out at a bad time. Making a list helps you remember what to do and when to do it. It also helps make sure that no tasks, even the minute ones, are forgotten. For example:

1. Check harvester chain tension.
2. Lubricate gun piston.
3. Check drive wheels for tightness and wobble.
4. Check main structure for loose parts or screws.
5. etc.

 # 302. TRICK: Check motor collar bearings between rounds to make sure the shaft does not wobble.

The "collar" bearings on the 12v motors are notorious for separating from the motor body. Once this happens, the motor is useless. The shaft will wobble, the gears will grind, and it will eventually stop working completely. In the photo we see an extreme example where the collar bearing has come completely off the body and slips down the shaft. Usually they will just become loose and wobbly before it gets this bad. You can check by trying to move the shaft by hand: if it moves laterally in relation to the motor, then the collar bearing is broken and the motor will work poorly and soon not at all.

 # 303. TRICK: Check the motor gearbox by seeing if the shaft turns without any resistance.

This is especially easy to do if you use direct drive. There should be resistance when you try to turn the 12v motor shaft. If there is not, this indicates that the gears inside the motor are stripped and the motor is now useless.

 # 304. TRICK: Check for servo motors about to fail by listening for a clicking sound when they are actuated.

The primary cause of failure for the servo motors is that the plastic gears inside get stripped either by over-rotating or by trying to lift too much load. Frequently, before they completely fail there is a tell-tale "clicking" sound when they actuate. Listen for that sound; if you hear it then you are close to a servo failure and you should replace it as soon as possible.

 # 305. TRICK: To quickly replace a burned out motor that is soldered, splice in a new pre-soldered motor rather than change connections at the motor controller.

Figuring out which wires go with which motor can be error prone and time consuming, and it is not terribly easy to make a good solid connection with the controller port under time pressure during a tournament. It is far simpler and less error prone to cut the wires near the burned out motor, strip the ends about 1 inch, and tightly twist together the new motor wires (red to red, black to black) then insulate these twist connections with some electrical tape. Use cable ties to take up any slack wire from the pre-soldered motor. When you have more time after the tournament you can trace the wires back to the controller and get rid of the excess.

Of course, if you use the plastic terminal connectors now available, there is no need to splice, you simply pop the connector off the bad motor and then back on the good one. This tip only applies to soldered motors.

Practicing for Judging

Judging is a supremely important aspect of every competition. It allows you to explain—in your own words— just how hard you and your team have worked to get this far. Some aspects of your robot may not be obvious from matches, and those unseen aspects might prove important for earning judged awards. Preparingyou're your judging sessions should not be an afterthought—it is just as important as building the robot and competing on the field. More than once our performance in the judging room has proved more important than our performance on the field. If your team's strengths lay in outreach activities and key learning experiences, be sure to explain that to the judges–it might make the difference.

 # 306. TIP: Take judging seriously, it must be practiced just like driving.

There are two ways to qualify for the World Championship: you can either win a major Championship tournament, or you can qualify by winning the Inspire Award and in some cases the Think Award. These awards are based both on your notebook and your performance during judging. By practicing your presentation skills, you increase your chances of qualifying by winning one of these awards. You can refine your message and speaking skills, and get across all the points you need to make to the judges.

We practice by holding mock judging sessions. In the 2009/10 season we had each of our teams practice judging several times during the course of the season. We did win several judged awards in scrimmages and a major award at the New Jersey State Championship. Team members improved dramatically with practice.

 # 307. TIP: Divide up the team into specialists to answer specific kinds of questions.

For example, the person who worked the most on the drive train might explain that part of the robot, the people involved in the harvester or scoring mechanism field those questions, etc. This is important because some judges will actually go around the room asking each person what function they performed on the team.

But how you designed and built the robot is not the only information you should be communicating. Don't forget about community service, driving, game strategy, the engineering notebook, fundraising and other crucial team functions. Divide up those topics among your teammates so they can practice delivering information on those topics.

 # 308. TIP: However, every team member should know a little about everything.

Although you may have designated topics for different students, sometimes you will run across a judge who will ask a specific question to a specific student. For that reason, it is important for every student to have at least some working knowledge of different aspects of the robot. However, if a judge asks you a question that is completely outside your area of responsibility, it's ok to hand off to the expert by saying something like: "Sarah worked on that part of the robot far more than I did, so Sarah would you like to take that question?"

 # 309. TIP: The entire team should speak. Don't leave people out.

Even if the judges do not specifically question every person, dividing up responsibilities is a good way to make sure everyone participates in the judging conversation. Many judges are looking for how every single member of the team contributed in a meaningful way. Not everyone will speak an equal amount, but everyone should say something useful. When you practice judging, make sure everyone speaks.

 # 310. TIP: Don't interrupt your team mate who is speaking, except to help him or her out.

If someone is answering a question try not to talk over them or interrupt them. If you have something to add, wait for a natural break in the conversation. The exception to this is if the person answering gets "stuck." For example, they forget the name of a certain part and are stuttering and searching for the right term, in that case go ahead and chime in to help your team mate out, but then let them continue to answer the question.

 # 311. TIP: Rehearse some "set pieces" in response to specific questions.

Some kinds of question get asked over and over again. For example, judges are likely to ask about your community outreach activities. For these kinds of questions, thinking about the response beforehand, and even practicing delivery of the answer, is a good idea. Especially when it comes to community outreach, if you team has done many things then practicing a "set piece" where you rattle off all the different activities is a good way to ensure you place the most important ones first, and that you don't forget any.

 # 312. TRICK: The judges are looking for passion. Be animated, confident, make eye contact, and smile!

Despite the prior TIP, it's important for you to avoid over-rehearsing. Rehearsing too much can make you sound like a robot! (Sorry.) Being spontaneous and showing real enthusiasm and passion for robotics is a sure way to win over the judges. And don't forget to smile!

 # 313. TRICK: Hit the most important points first. There is often very limited time. Keep answers short and to the point.

Using short, concise answers shows that the speakers are knowledgeable and allows for more time to explain everything about the team and robot. Rambling has the opposite effect. When you've answered the question, just stop! This is actually a hard thing for many people to learn but it is a crucial speaking skill. Practicing presentations during team meetings helps fine-tune presentation skills.

 # 314. TIP: Stress community involvement and outreach activities.

The judges love to hear how teams are spreading the word of FIRST. If your team did many outreach events, be sure to list all of them first, and then go into detail if asked or at the end if the "Is there anything else you would like to add?" question arises.

 # 315. TIP: Point out members who started in FLL.

This shows the judges the passion that those students have for FIRST robotics and shows your robotics program includes different levels of FIRST. Progression within FIRST and the intention to continue is a sign of commitment that judges like to see.

 # 316. TIP: Point out how you are helping start other FIRST teams.

Helping to spread FIRST to other schools and organizations in your area is a great way to demonstrate community involvement. If you have helped other teams to start up by advising them, offering to let them observe your team or use your field for practice, or you have helped run outreach events to inspire groups to start new teams, make sure you point this out to the judges. If you have not done things along these lines, then make them a part of your plans for outreach in the future!

Let's not forget: one way to help new teams is to contribute tips to this book! Email your tips to pj.ftc.tips@gmail.com, and if we use them you and your team will receive attribution in the acknowledgements as well as free copies!

Part IX: At the Tournament

Your team has spent long hours analyzing the game, prototyping, building, practicing and training. Now the day of your first tournament has arrived! This section will give you some ideas on how to handle the excitement, thrills and (sometimes) spills your robot is sure to encounter.

Types of Tournament

There are several different types of tournament and you may participate in all of them during your season. Most tips in this section will apply to any type of tournament, but there can be some exceptions. The types of tournament are:

- **Scrimmage**
  A scrimmage is an "unofficial" tournament usually held relatively early in the season, and its purpose is to practice against real competitors in the season's new game. Scrimmages may or may not include judged awards. Scrimmages normally will use exactly the same rules as the official tournaments, but sometimes there can be exceptions. For example, in New Jersey, scrimmages do not allow alliance captains to choose other alliance captains during alliance selection (more on this later). It is important to remember that scrimmages are also practices for referees and tournament officials, who are also learning a new game that has never been played before. Sometimes scrimmages even reveal flaws in the game, field elements, or rules that need to be clarified or even altered by the game committee before real tournaments take place.
- **Qualifier**
  In some regions your team must first take part in a Qualifier tournament before being allowed to compete in your regional championship. Rules differ on exactly how many teams attending each Qualifier will be admitted into the regional championship. For example, in the 2010/11 season in New Jersey there were three Qualifier tournaments and all three members of the winning alliance, the team hosting the qualifier, and every team that wins a

major judged award were all admitted to the New Jersey State Championship. The second place finalist alliance team and other award winners entered a lottery and had a chance of gaining a slot in the Championship.

- **Regional Championship**
  This is a tournament that may qualify your team for the World Championships. Regions are usually states, for example the New Jersey Championship, but can also be a large city (The New York City Championship) or even a country. Although rules vary on a regional basis, typically the captain team of the winning alliance and the Inspire award winning team gain an automatic berth in the World Championships. The winning alliance partner teams and sometimes the Think Award winner may enter a lottery system and have a percentage chance of getting invited to the World Championships.
- **World Championships**
  The highlight of the season is the World Championships. Only teams who have qualified by winning in a Regional Championship (or winning a major judged award there) are invited to the World Championships. In recent years the FTC World Championships were held at the Georgia Dome in Atlanta, but in the 2010/11 through 2012/13 seasons they were held in St. Louis, Missouri. In the 2009/10 season, only 100 teams out of about 1,300 were invited to this prestigious event.
- **Post-Season Events**
  Your region may host one or more post-season events. These events are just for fun, and give you one final opportunity to play the season's game with the robot you worked so hard to build. In some post-season events the rules of the game may be altered, for example the Monty Madness FTC post season event features the normal season game during qualifying rounds followed by an "extreme" version of the game for elimination rounds. Post-season events frequently do not have judged awards, but may instead have other kinds of award (most improved team, Gracious Professionalism award, etc.)

General Outline of a Tournament

Before diving into tournament tips, it's important to have an understanding of what happens at a tournament. If you are a new team, take a few moments to read this section. Knowing tournament terminology will help you understand the tips that follow.

While details will vary somewhat from one tournament to the next, the general sequence of what happens at a typical tournament is as follows:

1. **Arrival**
   Your team arrives at the tournament, unpacks your robot and supporting materials and enters the venue. Be sure to arrive at the earliest allowed time; this gives you the ability to recover if you have any kind of difficulty.
2. **Registration**
   Your coach or a team member goes to a registration desk and reports that your team has arrived. At this time you will be asked to hand in registration documentation such as FIRST release forms, and you may also be asked to hand in your engineering notebook. You will probably receive an information packet that contains the day's schedule of events, information on food concessions, driver buttons to be worn by your team's drivers, and other useful materials. Don't lose this packet!
3. **Setup**
   Your team goes to the "pit area" where typically you will find a table with your team's number displayed. You must wear safety glasses at all times in the pits. Your team decorates your pit area, unpacks the robot and lap top, etc. You should immediately check to make sure your robot is in running order in case it was damaged during travel. You typically will have a power cord in your area that you may use to charge batteries, laptops, etc.
4. **Inspections**
   As soon as your robot is up and running, you should get in line for hardware and software inspections, which can sometimes take significant time. When you have passed inspections, you will receive a sticker or some other kind of mark on your robot.
5. **Judging**
   Sometimes this occurs as part of inspections, but more often it is a separate item and in some cases can take place throughout the early part of the tournament, even after rounds have started. In some tournaments you will receive a judging time slot after you register, and you are expected to show up at the judging room at that time. At other tournaments you will proceed to judging right after inspections are complete on a first come first served basis. Judging is used to select teams who win awards and it is an important part of the tournament, so be sure to keep track of your time slot. All team

members should attend the judging session. Mentors typically may also attend but must keep strictly silent; only the students should speak.

6. **Opening Ceremony**
   This typically occurs an hour or two after the doors open. Judging and inspections may not be complete before the opening ceremony, they are usually put on hold so everyone can attend. Usually the tournament organizers will give a brief speech, and there may be invited guests who speak briefly as well. The National Anthem is often performed, and if there are teams from other countries attending the tournament, their national anthem(s) may also be a part of the ceremony.
7. **Practice Rounds**
   Some tournaments have time for some "practice" rounds before real games are played. Frequently, tournament directors will only allow teams who have completed inspections to participate in practice rounds. It is greatly to your advantage to get your inspections done as quickly as possible and participate in these practice rounds, as this is a chance to find out if the field conditions affect your robot's performance, requiring you to make adjustments before real games are played. (For example, minor field variations may affect autonomous mode, light sensors may face conditions not present on your practice field, etc.)
8. **Driver Meeting**
   This usually occurs shortly after the opening ceremony and the drivers on the team will receive a briefing from the tournament directors and perhaps referees. There can be important information at the driver meeting so be sure your drivers attend.
9. **Qualifying Rounds ("Qualifiers")**
   Schedules showing which teams play each other are posted and qualifying rounds begin. (Do not confuse the qualifying rounds of a tournament with the concept of a Qualifier Tournament!) At most regional championships there are four or five qualifying rounds. Your team is paired with a random partner robot, and you face randomly paired opponents during qualifying rounds. The purpose of qualifying rounds is to identify the top few teams, who will become captain teams for the elimination rounds later.
10. **Lunch Break**
    Sometime during qualifiers there may be a lunch break, although in our experience at most tournaments it ends up being reduced or eliminated due to time constraints! So team members need to eat between rounds when they can in many cases.

11. **Alliance Selection**
    After all qualifier rounds are completed, the teams are ranked in "seed order." The team with the best record is "first seed," second best is "second seed," etc. There are specific tie-break rules to place teams in the case they have an equal number of wins, see elsewhere in this chapter for a discussion of the tie-break system, called "ranking points." The top few teams (typically four per division) become "captain teams." The captain teams select partner robots for the elimination rounds.
12. **Elimination Rounds**
    Each alliance, composed of an alliance captain team plus one or two alliance partner teams (depending on how large the tournament is) now face each other. The first seed alliance faces the fourth seed, and the second seed faces the third seed in the semi-finals. They play best two of three games, with the losing alliance being eliminated. The two winning alliances in the semi-finals then face each other in the finals, again in a best two of three games format. In large tournaments there may be two divisions, and the winners of each division would then face each other in a grand final.
13. **Awards and Closing Ceremony**
    Immediately after the final (or grand final if there are two divisions) is complete, an awards ceremony is held. The judged awards are announced at this time, and awards are distributed for the winning alliance and the finalist alliance.

Each tournament is a little different, but these basic phases will occur in some form at all tournaments. Scrimmages may have some special rules, such as restrictions against captain teams choosing each other for alliance partners.

Planning Your Season

 # 317. TIP: Attend at least one scrimmage. Two would be better.

Nothing substitutes for practice under real conditions. The closest you can come to this without playing an actual championship is to attend one or more scrimmages. If your area does not offer scrimmage opportunities then try to find some in neighboring regions, even if you have to do some traveling. If even that is not possible, try to find teams in your area and organize a miniature tournament.

318. TIP: *Try to participate in at least two Championship level tournaments, even if you have to go to a neighboring state.*

Even the best team with the best robot may not win all the time. A thousand things can go wrong from dropped wireless connections to fried motor controllers. Giving your team two opportunities to qualify for the World Championships is obviously better than only having a single shot.

This is easier in the northeast states that are smaller and closer together, of course, and geographic realities or budget constraints may make this impractical for your team. But at least explore the possibility of attending more than one Championship tournament.

319. *Register for tournaments as early as you can in the season. Pay attention to qualifier requirements. Some tournaments fill up quickly.*

You don't want to be stuck without an opportunity to compete in your regional Championship tournament. Pay close attention to the FIRST web site and the web site for your local chapter of FIRST for complete details on how to qualify to compete in your regional Championship. Register as early as you can to avoid getting caught without a spot.

General Tournament Tips

320. TIP: *The mentors and coaches should each have a master list of cell phone numbers for everyone attending.*

Nothing is more frustrating than not being able to locate your drive team or get information from your scouts before a critical round of play.

321. TIP: *Always inform your coach/mentor when you are leaving the pit area to eat, watch a match, etc.*

We've had plenty of cases where we desperately needed someone at our pit, only to find out later they were off eating lunch somewhere!

322. *Use Twitter or similar services for reporting results to parents back home, and the team themselves.*

Twitter (www.twitter.com) is a great way to broadcast out the results of rounds or other tournament status information for parents and others who couldn't attend. It's even useful in a larger tournament for communicating between team members. Be careful that you have an appropriate text messaging plan that doesn't have high per message charges though!

 # 323. TIP: Always put your DRIVER/COACH designation buttons in the same place when not being worn by the drive team.

We normally put them right in the emergency toolkit box that stays on our rolling cart. Stress to the drive team that they should either be wearing the buttons or have the buttons in the designated place, never anywhere else! At some tournaments they strictly enforce that only students with the proper driver/coach buttons are allowed on the field, and they may not give you a duplicate set if you have lost them.

 # 324. TIP: When starting a match, use CONTROL-S on your keyboard instead of clicking "Start Match" in the Field Control System.

This key combination is safer than clicking "Start Match" in the FCS. Because it doesn't rely on the mouse pointer to start the match, you can press these keys without looking at the screen on your laptop. It's also a lot harder to accidentally start the robot if your mouse isn't hovering over the "Start Match" button before a game. Clicking CONTROL-S will start a match if one isn't active, start tele-op if you hit it in the delay after autonomous mode, and stop the match if you press it during tele-op or autonomous.

NOTE: This TIP was obsolete in the 2010/11 and the 2012/13 seasons because FTC moved back to a system where your drive team does not use their own laptop, but rather a field-supplied control system is used.

 # 325. TIP: Calibrate any light sensors you use on the competition field as soon as you get to the tournament. Pay attention to windows and sunlight conditions that might change lighting levels throughout the day.

One of the first things you should do after arriving at the tournament is calibrate any light sensors you may be using. If possible, do this on the actual field where competitions will occur.

Pay attention to windows. We attended a scrimmage that took place in an atrium. Light levels changed throughout the day as the sun rose in the sky or as clouds passed by. Sometimes there would be bright squares of light in some parts of the field and shadow in others! We actually had to construct a plastic "shade" to reduce these effects, which were wreaking havoc with our autonomous program.

326. TIP: *Make sure at least one student or mentor video tapes your rounds of play.*

Most people have digital video cameras that can do a great job of recording your matches. Besides being a great memento of your season, the recordings can be used to analyze how your strategy and hardware worked, which can be crucial for designing improvements or better preparing your driver team for the next tournament. Designate either a student or mentor as the team videographer and try to film every round of play.

327. TIP: *After qualifiers, video tape the score board long enough to record all the results.*

This will help you record the final standings of all the teams at the tournament for later analysis. Frequently tournaments do not hand out hard copies of the final results.

328. TIP: *A monopod helps you get good video recordings of your rounds.*

A monopod is an expandable pole that screws into your video camera. They are usually six to seven feet long. By using a monopod, you can hoist your video camera high in the air, over the heads of bystanders, to get much better pictures of your tournament play. A monopod is much lighter and more compact than a tripod, allowing for easy use in a busy tournament. You can find inexpensive ones that do the job for about $15 to $30 online at places like Amazon.com or wherever camera equipment is sold. Alternatively, ask if any parents already own one.

Preparing your Laptop for the Tournament

NOTE: The tips in this section may be obsolete for tournament play in the 2010/11 season, because early indications are that FTC will not be using each team's personal laptop for competition. However, these tips do apply to setting up your laptop for practicing on your own field.

329. TRICK: *Make sure that your laptop virus scans are disabled during game play.*

We've seen cases of a virus scan deciding to run right as a round began, causing lag in controls or even lost Bluetooth connections. Other background programs may also cause the same issue. Even an email program suddenly deciding to go out and download huge files can cause problems, and you can cure that by making sure your email program is not running, or simply not installing one at all.

 # 330. TRICK: Make sure laptop power saver settings are set so the laptop never goes into hibernation or sleep mode.

We've also had cases where the laptop used to control the robot decided to go to sleep while the team was queued up for a round. Some laptops also automatically sleep or hibernate when the cover is closed, or will turn off disk drives after a few minutes of inactivity. It is best to go into the System settings and turn off all such power saver features during a tournament.

 # 331. TIP: Reboot your laptop and your NXT Brick every few rounds of play.

Computer systems just work better if they are occasionally rebooted. This "cleans up" memory and zombie processes. Be sure there is plenty of time to the next round before rebooting, however. This also goes for the NXT brick. We have found that after several rounds of play the chances of an NXT brick crash start to increase dramatically if you do not reboot it.

 # 332. TRICK: Do not load any unnecessary software on your laptop, especially as you approach the tournament date.

Every time you load software on a computer, you risk interfering with the other programs running there. Many programs do nasty things like put background processes in your startup menu, sapping computing power constantly, even when the program in question is not being used. Keep your tournament laptop clean and running fast by not installing anything you don't really need.

Packing and Preparing for the Trip

 # 333. TIP: have at least two, and preferably three, 12v batteries for each robot, and at least two rechargeable NXT batteries. Have at least two of each kind of battery charger.

Because autonomous programs and even tele-op performance is greatly influenced by battery level, you will want to always use a fresh, topped-off battery for each round of play. We have found that three batteries per robot is almost always sufficient to ensure you can start each round at full power. Also, chargers do go bad, so make sure you have at least two of each kind of charger (NXT and 12v).

 # 334. TIP: Pack up a small box with an assortment of metal parts, brackets, channels, flats, etc., as well as pieces of plastic or sheet metal, for repairs.

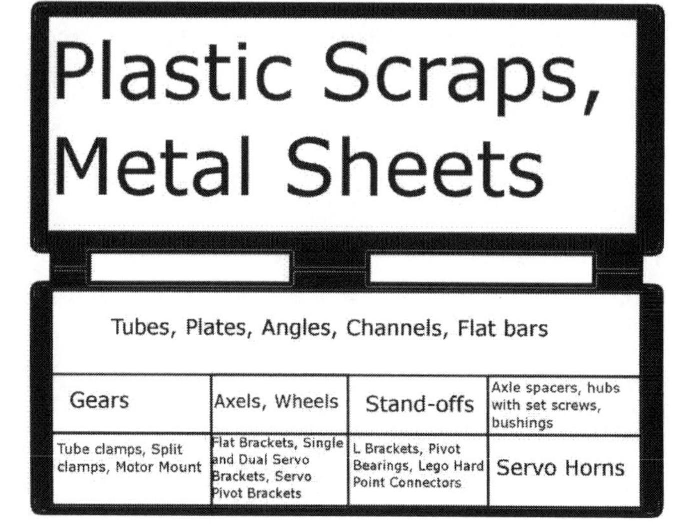

You don't need to bring huge quantities of repair materials to the tournament, but an assortment of metal beams, flats, angles, brackets, and some chunks of allowed plastic or sheet metal can come in very handy for minor repairs.

 # 335. TIP: Bring spare electrical components.

Motors, servos, motor controllers, servo controllers, and sensors. All of them can be damaged during rough play and may need to be quickly replaced. Bring at least one spare for each major part if your budget allows.

 # 336. TIP: You don't need to bring every tool you have, but a basic assortment, including a hand drill and a hack saw, are necessary for common repairs.

Your emergency tool box should have most of what you need, but larger tools like a hacksaw, electric drill, etc., probably won't fit in it. Bring along a box of these larger items.

 # 337. TRICK: Get a rolling cart for your robot that can be easily disassembled.

Many such carts are available for under $100. Try to find one that can be broken down and assembled without tools for easy transport. Plastic is usually lighter than metal!

 # 338. TRICK: Modify your cart for added convenience.

We used a cart that was easy to break down for transport, but found at one tournament we had to carry it up a flight of stairs. The tubes holding it together popped apart, nearly spilling our robot on the floor. After that we modified the cart by drilling some holes in strategic places and using bungee cords on each end to securely hold it together. We also added a strip of metal on one side to form a lip so the robot could not possibly roll off.

 # 339. TRICK: Create an emergency repair kit that stays with your cart during rounds.

Organize a small plastic case that contains an assortment of tools, spare nuts and screws, cable ties, rubber bands, and other materials that can be useful for quick repairs.

 # 340. TIP: Customize your emergency repair kit based on your robot's specific design.

Some items like cable ties are universally useful for repairs; others may depend on your specific robot. For example, if you robot is heavily reliant on rubber bands it is important to include rubber bands in your kit. In the 2010/11 season, one of our teams had a robot that used a plastic coated wire rope to hoist a dispenser mechanism, so they packed a few feet of wire rope in their kit. Our other teams didn't need that because the design of their robot didn't require wire rope.

 # 341. TIP: Keep the emergency kit complete, yet light.

While completeness is important, having too many items in your emergency kit is also not good. Make sure you have one of each sized hex key, but there is no need to have multiple units of each tool, for example. A cluttered toolkit just makes it hard to find things.

 # 342. TIP: Bring a printer/copier to the match.

If budget allows, a printer/copier can be very useful for duplicating match schedules (usually in short supply) and printing scouting reports.

 # 343. TIP: Print out self-stick labels with your team logo and affix them to batteries, chargers, repair kits, spare part bins, large tools like drills, etc.

You may loan out these items to other teams, or your pit table may be right next to several others. It can get confusing to sort out who owns which items if they are not marked.

Traveling by Air with Your Robot

 # 344. TRICK: Obtain 19" cube boxes and 18" cube boxes, put one inside the other with packing material filling the spaces. Carefully pack your robot inside.

We used this technique to get two robots to the World Championships in Atlanta and it worked very well. Both robots made it through as checked airline baggage without a scratch. You can buy cube boxes at online stores like www.uline.com but unfortunately you have to buy them in quantity 10. Perhaps you could share your extras with other teams in your area, or just save them for succeeding years. You may be able to find 18" cube boxes in single unit quantities at Staples, Office Max, or other office supply stores.

For air travel, it is very important to understand the checked baggage size and weight restrictions. At this writing, many airlines allow baggage whose length plus width plus height is 62 inches or less. The 19" cube box falls under this limit even if it bulges out a bit after being stuffed with packing material, which is why we chose it. But you must check your airline; they are changing the rules all the time. Weight restrictions are equally important to understand. Keep your boxed up robot at least half a pound to a pound under the weight limit in case your scale does not match the one at the airport.

 # 345. TRICK: Remove the wheels before packing.

This is especially a good idea if you use direct drive, because hard bumps will stress the motor shafts directly. The bearing that supports the shaft is notorious for breaking under strain. The last thing you need is to have to swap motors as soon as you arrive at the tournament. It doesn't take long to put wheels back on however, as long as you place the hub is between the wheel and the chassis it only takes a couple of screws that are easily accessible. Removing the wheels also gains an inch or so of extra space that can be used for packing.

 # 346. TRICK: Remove the battery and NXT brick before packing.

We prefer to carry on the NXT brick with the laptop. Then 12v batteries should be wrapped in bubble wrap or some other material, then taped up to protect them from getting punctured, and most likely will have to be in checked baggage (we put them in the box with the robot). You almost certainly would have trouble trying to carry on a 12v battery. The NXT battery just goes with the brick. It is no larger than other consumer devices like PDAs and laptop batteries.

 # 347. TRICK: Mark the robot boxes with your team logo and an explanation letter to TSA inspectors.

Print out your team logo and tape it to the box for identification purposes. Label the box with an adult mentor's name and address (one who is coming on the trip of course). FIRST published a sample letter in the 2009/10 season to affix to the box alerting the Transportation Safety Administration (TSA, the organization who inspects baggage at airports) that the contents was bound for a robotics competition and requesting that they be careful during inspection not to damage the robot.

Pit and Robot Decorations

 # 348. TIP: A six foot banner is a cost effective pit decoration.

For relatively little money you can have a color banner printed on vinyl or a similar substrate. This can be simply duct-taped to your table (in front), or you can use PVC pipes to easily create a stand to support it over your table.

 # 349. TIP: Print your banner two-sided.

It usually doesn't cost much more than one sided, and at some tournaments the pits are arranged such that both sides will be visible.

 # 350. TIP: Use self-stick labels to affix team numbers to the robot.

There are all kinds of creative ways to make your number. If you want to get it laser cut from aluminum, or fancifully paint it on, more power to you! But from experience we know the numbers and decorations take a beating and get scratched up and sometimes shredded from intense battles on the field. It's nice to just be able to stick a new number on top of a ripped one to make your robot look good again.

 # 351. TIP: Use photo quality paper to print the team name and sponsor logos to attach to the robot.

This is a great way to make your robot look really nice in a simple, relatively inexpensive way. You are allowed to use tape to affix decorations, as long as the tape and decorative materials have no "function" or structural value. To ensure you don't run afoul of this rule, it's best to tape such decorations to legal materials like plastic or aluminum sheet.

 # 352. TIP: Have a "theme" for your team that carries through to the pit decorations, team uniforms, and robot decorations.

Some teams are very memorable because they do a great job of carrying a theme throughout their pit area, robot, and even team uniforms. For example, "Team Tiki" (2859) uses a tropical island theme with brightly colored, elaborate costumes (including flowers and head gear), and can often be seen marching through tournaments bearing an island carving and chanting "Tiki, Tiki, Tiki." Who could forget that!

But don't think such thematic decorations need be expensive. "Team Duct Tape," (2845) creates clever decorations for their booth entirely out of, you guessed it, duct tape! A memorable theme at a budget price.

Impressing the Judges

See the prior section on "Practicing for Judging" in Part VIII for tips on this topic. At the real tournament, simply use all the skills you have practiced!

Passing Inspections

 # 353. TIP: Use pre-tournament inspection check lists to check out your robot before even arriving at the tournament.

In recent seasons an inspection checklist is made available to teams via email before the tournament even starts. Make sure you are familiar with this checklist and ensure your robot meets all its requirements to the best of your ability.

 # 354. TIP: Get into inspection lines as early as possible upon arriving at the tournament.

As soon as you arrive, get your robot running and connected to your laptop, then get into line for inspections. This gives you maximum time to work out any problems that might be found.

 # 355. TIP: Be very careful that last minute robot modifications do not bring you out of compliance.

Adding a last minute piece of shielding just might put you over the size limit by a sixteenth of an inch. Be careful about any modification made just before the tournament that could cause you to fail inspection.

 # 356. TIP: Have your cut sheets and material receipts available for the inspectors.

Prior TIPS have suggested ways of creating "cut sheets" that show exactly how each part of your robot was cut from sheet metal, sheet plastic, and other materials allowed by the season's rules. These are crucial for proving that you are within material limitations if questioned by hardware inspectors. In addition, keeping receipts for all these extra materials can help prove they are the correct thickness and substance. For example, in some cases plastic or sheet metal may slightly vary in thickness from its "nominal" thickness. In other words, 1/16" thick aluminum might really be 1/15" thick or 1/17" thick when it comes from the manufacturer. By showing that you bought the correct "nominal" size you should be able to successfully get through an inspection where the inspector uses calipers to measure the thickness and finds it is slightly off.

 # 357. TIP: If you fail inspection, calmly correct the problem and get back into line as soon as possible.

Don't panic. We've failed inspection a couple of times due to minor issues. If you got into line early to begin with and you work quickly, you should be able to get through on your second attempt. In practice, the great majority of robots end up passing inspection. The only exception to this that we've seen is those cases where a robot flagrantly violated materials restrictions, to such an extent that it was not possible to come back into compliance. This is why several other TIPS in this book caution you to understand and be certain you are complying with materials restrictions!

Pit Crew

 # 358. TIP: Keep the pit tools and emergency repair materials organized.

Know where everything is so you don't waste time! You need spare motors and servos, motor controllers, an assortment of metal and screws, spacers, cable ties, etc. all at your fingertips.

 # 359. TIP: During elimination rounds remember that you can call a repair time out.

Be sure to ask about this rule at your specific tournament, but in most tournaments we've been in they do allow a time out for a quick repair at least during elimination rounds. The amount of time allowed may vary but is typically a few minutes, which can be enough time for an emergency repair. They may be less flexible during qualifier rounds where you have more time between games.

 # 360. TIP: Cable ties are great for quick repairs.

There are many ills that can be rapidly repaired by clever use of cable ties. Cable ties can often be applied to hold on parts of your robot in a few seconds when doing a "real" repair with screws would take far longer.

Scouting

When you think about FTC competitions, what first comes to mind is the robot and drive team's performance on the field. But behind the drivers and the robot are important support activities. Scouting is one of those important activities. We have had many cases where a round was won or lost due to the quality of scouting information. Scouting is supremely important in those very close matches where seemingly minor strategy decisions determine whether you win or lose. Scouting has to be practiced and taken seriously in order to reach the top of major competitions.

 # 361. TIP: Create a "Round by Round Scouting Report" specific to the season's game that reports, by team number, the capabilities and effectiveness of each robot in a round of play.

Creating a game report for each scrimmage and tournament that you attend is crucial for success. It allows you to track which robot is good at which task and how effective they are at the task. When your team gets to the elimination rounds of the event, it is good to know what strategy to use against the teams you are up against and the game reports help tremendously. Reports are also useful for alliance selection if you are a captain team.

The image shows a part of our "Round by Round Scouting Report" from a recent season. There are columns for each major part of the game (this will be specific to each year's game). A scout fills in this report each time he or she observes a round, tracking each robot by team number. This was created using a spreadsheet program. Each sheet could track up to eight rounds of play; only a couple rounds are shown here.

	A	B	C	D	E	F	G	H	I	J	K	L	M	N	O	P	Q	R
1	ROUND BY ROUND SCOUTING REPORT																	
2	Round	Team #		Low Goal	High Goal	Drop Balls	Defense		Low Goal	High Goal	Off Field	Doubler	Defense		Score	Problems	Impression	Notes
3	Rd	Team #		LG	HG	DB	D		LG	HG	OF	DB	D		Score	Prob	1-10	Notes
4			Autonomous					Tele-Op						Results				
5																		
6																		
7																		
8	Rd	Team #		LG	HG	DB	D		LG	HG	OF	DB	D		Score	Prob	1-10	Notes
9			Autonomous					Tele-Op						Results				
10																		
11																		
12																		
13		Team #			HG	DB	D				OF	DB					1-10	

 # 362. TIP: Create a separate report that just summarizes the best teams that scouts have observed.

The round by round game report allows your scouts to gather information about each potential opponent and partner robot. However, you also need a way to quickly see which robots are going to be the most trouble (or the best partners during alliance selection.) The "Best Teams" report summarizes information only about the

	A	B	C	D	E	F	G	H	I	J	K	L
1	Best Teams (Impression is a scale of 1 to 10)											
2	Team #	Auto. HG	TeleOp HG	Off Field	Doubler	Defense	Heavy	Sturdy	Driver Ability	Strategy	Impression	Team #
3	Team #	Auto. HG	TeleOp HG	Off Field	Doubler	Defense	Heavy	Sturdy	Driver Ability	Strategy	Impression	Team #
4	Team #	Auto. HG	TeleOp HG	Off Field	Doubler	Defense	Heavy	Sturdy	Driver Ability	Strategy	Impression	Team #
5	Team #	Auto. HG	TeleOp HG	Off Field	Doubler	Defense	Heavy	Sturdy	Driver Ability	Strategy	Impression	Team #
	Team #	Auto. HG	TeleOp HG	Off Field		Defense	Heavy	Sturdy	Driv	trategy	Impress	am #

most dangerous opponents that the scouts have observed. Generally, any robot that has a reasonable chance of beating your team one on one should be on this report. The information on this report may change as you observe each opponent in multiple rounds of play.

Before each round, consult this report to see if you are up against any of these dangerous opponents or to see if your partner robot is on the list. This can help you map out strategy for your most difficult rounds.

 # 363. TIP: Scouts should also walk around the pits and observe possible capabilities of opponents they have not seen play yet.

Scouting reports won't help you in your first round or two of play, since the scouts have probably not seen all your opponents in action yet. For this reason, scouts should be out looking at robots in the pits even before the first round of play. Scouts can also observe practice matches or how opponents work on the practice fields that many tournaments provide.

 # 364. STRATEGY: Scouts should look at the upcoming rounds for your team and a "scouting coordinator" should report to the drive team and coaches the capabilities of both partner robot and opponents ahead of each round of play.

Look ahead. Scouting from an early point can mean the difference between winning and losing for your team. If you scout early enough, many teams will be known to you and your teammates. Strategies for rounds are based off these scouting reports and can determine which team will be victorious.

You may want to designate one person as the "scouting coordinator" who will receive the round by round game reports from other scouts, transfer information to the "Best Teams" list and be responsible for briefing the drive team before every round of play. The scouting coordinator will also analyze the match pairings when they come out and will focus their efforts on obtaining information about your team's actual opponents. However, all robots in your division must be scouted even if you will not play them during qualifier rounds, because you will need to know who to choose for an alliance partner if you become a captain team later.

Also, don't forget to scout your alliance partners as well as opponents. Many teams over-estimate their robot's capabilities so you want to observe firsthand how reliable they are. It is common for a good team to lose a round because their partner claims to have a great autonomous capability that requires all the initial scoring elements, when in fact it is too unreliable to score.

Driver Team

 # 365. TIP: Don't panic before a round. Take a deep breath. Relax.

No matter how much you practice, when you get to a crucial round against real opponents you may get nervous. If you feel "butterflies" in your stomach, that's not necessarily bad, it may keep you sharp. Just take a few deep breaths and remind yourself that you've done everything you can to prepare for this moment, that you are ready. Then just do your best; that's all anyone can ask!

 # 366. TIP: Always be extremely respectful toward the referees, your opponents, your partner team, and your own teammates.

Like any sport, there may be times when you disagree with a ruling by a referee. It is important to keep in mind that the referees are volunteers who are giving up their free time to help run the tournament. If you disagree with a call the referee made, and you won the game anyway, it may be best to just let it go. If you lost the game because of what you considered an incorrect call, you may respectfully explain your position to the referee in a calm, factual manner. You should understand, however, that it is extremely rare for a referee to reverse a call and he or she is very likely to let the original ruling stand, perhaps with an explanation to you and perhaps not. Either way, thank the referee for listening to your concerns and then focus on your next round.

You also should be respectful to your opponents, even if you disagree with the way they played or perceive that they benefited from a referee decision that you disagree with.

You must remain respectful of your partner robot's team even if you feel they made mistakes that cost you the game. You should respectfully explain to them what you think they should have done differently in the spirit of helping them improve in the future.

This of course also applies to your own teammates who may have made mistakes. Teammates (and sometimes even parents) need to realize that losing is a part of competition and things sometimes do not go as planned. It's a lot harder to control these robots and make split second decisions during a game than it looks. Don't blame your teammates for the loss, but do try to explain to them how they could improve under the same conditions in the future.

These constructive ways of handling a loss are sometimes called "good sportsmanship." In FIRST robotics they go by a different name: Gracious Professionalism.

Game Strategy

Every year's game is different, so this section is necessarily going to have to be general. Review these general tips to ensure they apply to whatever game you face before employing them.

 # 367. TRICK: To turn an opponent away from their scoring or harvesting position, it is better to ram them in their robot's corner rather than flat on between their wheel bases.

It is often more effective to ram an opponent on one of their robot's corners to turn them away from a scoring position rather than making full contact along one side. In the diagram, ramming along a full side does little to move the opponent in most cases because you are going against the "grain" of all of their wheels. They may only move slightly laterally, and may still be able to score (or harvest, or whatever they are trying to do that you want to stop!) On the other hand, if you ram them in the corner then their robot will much more easily pivot and rotate out of position. You are only fighting half as many wheels, turning the other wheels into a fulcrum around which the robot will rotate. This is especially effective against robots that have a short wheel base, they will often turn easily.

Be aware that overly aggressive ramming may cause the referees to penalize your team, so be sure to consult the game rules. In past seasons some contact has been permitted to achieve strategic goals as long as you are not seen to be actively trying to damage your opponent.

Also be careful not to violate "pinning" rules. If you ram an opponent in such a way that you wedge them against field elements and prevent them from being able to move for more than a few seconds, you may be penalized for "pinning." To avoid getting called for pinning, it is best to back off a couple of feet after a successful ram to show that the opponent can still move. Backing off also has the advantage that if they reposition again you have room to gain momentum for another ram attempt to turn them yet again.

 # 368. STRATEGY: Occupying the opponent's preferred scoring position before they get there is often easier than trying to push them away from that position later.

A good example of this is the 2009/10 season, when many robots used a strategy of gathering many game elements then shooting them into off-field goals in the last 30 seconds of the game. Most of the robots using this strategy had a fairly narrow set of positions from which they could accurately hit the off field goal. If they were able to get to that position, however, they were hard to beat.

We found that by getting there first and occupying their "sweet spot" before they could, we could make. But if they made it to that desired spot before us, it was nearly impossible to dislodge them without getting called for "pinning."

So the strategy that was usually successful to defeat this kind of opponent was to outscore them in autonomous and the start of tele-op, then (once we had a lead) head straight for their desired field position and just park there, sometimes for as much as a full minute before the endgame. But to work, we had to get there before they did and resist the urge to move out of position.

 # 369. TIP: Do not use excessive force.

While it is normal for there to be some pushing, shoving, and jockeying for positions out there, you may be called for a penalty if your driving is interpreted by the referees as overly aggressive. For example, if you are going up against a robot half your robot's size and you ram it at full speed, perhaps causing damage, when a simple nudge would have achieved your strategic goal, you should not be too surprised when a ref calls a penalty or even disqualifies you. Try to use the minimum force necessary to achieve your objective. You should never give even the appearance of trying to purposely damage your opponent's robot. That kind of behavior is not within the guidelines of "Gracious Professionalism."

 # 370. TRICK: The coach member of the driver team should keep score on a pad of paper in order to make end game decisions.

The coach on the driver team has a crucial role to play. He or she must keep a view of the entire field and game situation—including an estimate of the current score—and advise the driver and gunner. The driver and gunner are usually very focused on driving the robot and manipulating its harvesting, scoring and other effectors, and therefore may not be aware of things happening on the other side of the field.

In addition, the driver and gunner may not have a good idea of where point totals stand at any given time. Having an accurate estimate of the current score is crucial for making strategy decisions.

For example, in the 2010/11 season, teams could score points during the last 10 seconds of the game by balancing one or more robots and rolling goals on a bridge. Generally speaking, it was easier to balance a robot alone rather than try to drag a heavy rolling goal up with the robot.

We saw many cases of teams (including sometimes our own teams!) waste time by dragging up a rolling goal when all they really needed to win was just the robot without the goal. In some cases, these bad choices actually resulted in losing a round.

In one instance, one of our teams was ahead 17 points going in to the endgame period. The opposing alliance did not have the capability to drag a rolling goal and they were going for a double-robot balance for 20 points at most. Our partner robot was dead so could not balance with us. However, the mathematics of the situation was that all we needed was to balance our own robot to guarantee a victory by 7 points. Our team had not been keeping close track of the score, and so they attempted to drag a rolling goal up onto the balance bridge unnecessarily. The rolling goal got stuck on the lip of the bridge, and they failed to balance at all, losing the match by 3 points rather than winning by 7.

As an aid, the coach should bring a pad of paper or clip board, and should compute the scores as they occur. For example, after autonomous write down scores, then track scores into each kind of goal by each team. At any given moment the coach should be able to tell the drive team a reasonably accurate estimate of how many points ahead or behind their alliance stands. This is the only way the team will know how to make a tough decision on whether to take a risk to score more points or whether to play it safe.

There are cases where it is difficult to know the exact score. For example, in the 2010/11 season it was not always clear after autonomous whether the referees would rule a robot to have successfully scored points by crossing the center line of the field. It was also not always easy to see the "low goal" and so there could be points "hiding" out of sight of the coach. In cases like this the coach simply has to take his or her best guess, or write that a certain number of points are "unknown." Even when knowledge is not perfect, a rough estimate is better than none at all.

 # 371. TIP: Do not unnecessarily jeopardize points already scored.

In the 2010/11 season your alliance could score points by dragging a rolling goal up onto a balance bridge. However, that rolling goal could contain batons that had already been scored and so by dragging it around you had to risk tipping over your own goal and losing those points. (Only the state of the batons at the end of the match counted, if the batons were not in the goal when the buzzer sounded they did not result in a score.)

Each alliance had two rolling goals. We saw many cases where a team had a choice of either dragging a goal that contained no batons, or dragging a goal that contained batons, and they inexplicably chose to drag a goal that contained batons. This unnecessarily risked those points that had already been put into scoring position. At the World Championships we actually saw a team spill their own goal that contained six or seven batons, losing a match that appeared to be an easy win.

Drivers must be drilled on points like this ahead of time, and the driver team's coach should also remind drivers that, when possible, they should make choices that do not risk already-scored game elements.

Everyone on the team should analyze each new season game looking for this kind of situation.

 # 372. TIP: Take robot-to-FMS communication failure into account when choosing strategy.

The Samantha WiFi module was introduced in the 2010/11 season and resulted in a very high rate of communication failures between robots and the field management system. There were many different sources of problem with the new system (power wiring, USB cable mechanical stress relief, site-specific WiFi environment, and others). By the end of the 2010/11 season, some of these issues had been resolved and fixes published, but it appeared that others still remained, and it is unclear what level of communication failures will occur in future seasons.

The problem seemed to be more severe at some competitions than at others, but it seemed to affect all competitions in at least some cases.

Unfortunately this is just a fact of life in the FTC competitions, and teams must design their strategies so to minimize the impact of these communication failures.

This kind of problem forced teams to think to themselves, "what if I were to lose communications right now?" and to control their robot accordingly. A good example in the 2010/11 season was that in the final 30 seconds of the match (the end-game) there was an enormous penalty, 40 points, for touching the opponent's balance bridge. We had a case at one tournament where very early in the teleop period one of our robots lost communications while crossing over the opponent's bridge. They could no longer move once the communications problem began, so they were still parked there with 30 seconds remaining in the game, and the referees hit them with the dreaded 40 point penalty, costing them the game.

For the rest of that tournament (which had extremely severe communications issues all day) our drivers avoided crossing the opponent's bridge *at any time.* Even with a full minute or more remaining, they treated the bridge as "radioactive" and avoided getting near it. They just could not guarantee from one moment to the next that the robot would remain under their control, so this was the best strategy given the circumstances.

Another example is that a robot could "capture" a rolling goal only in the last 30 seconds in order to balance with it on a bridge. However, the robot was allowed to balance by itself at any time during the game. In a situation where it appears unlikely that you would need more than a single robot's balance to win a game, your best strategy would be to balance as early as possible in the tele-op phase (forgoing the rolling goal since it could only be captured in the last 30 seconds when you may no longer have communications), that way you score the balance points even if you later lose communications.

A third way the communications problems affected strategy is that these problems put an even greater emphasis on autonomous mode. Because you never knew whether you would actually be able to control your robot during tele-op (or at least for all of tele-op) that meant those points you scored in autonomous might very well be the only ones you scored at all.

The communications issues were greatly reduced in the 2011/12 and 2012/13 seasons, but your team should always brainstorm ways to alter your strategy to take into account the possibility that at any moment you could lose communications either temporarily or permanently. You may not need those strategies at some tournaments that have fewer problems, but you are well advised to have such contingencies prepared in case you find yourself in a tournament with serious issues.

Qualifying Rounds

 # 373. STRATEGY: Pay very close attention to ranking points.

Many teams do not properly take into consideration the "ranking point" rules at tournaments, and this sometimes means the difference between winning and losing at the championships.

At the tournament, it is not just the number of wins you achieve that determines your final rank after qualifying rounds. In most tournaments, there will be ties in the top positions between teams that have equal numbers of wins. For those ties to break in your favor, you need good "ranking points."

What is a ranking point? The concept is very simple: your ranking points are the total number of points scored by the losing team in each match. The more points the losing team scores, the better your tie-breaks when it comes to determining which teams are captains. (Of course, you are going to try not to be the losing team!)

The strategy that many teams overlook is to actually score on their opponent's behalf when they are far ahead and likely to win by a large margin. This is something you should actually drill your drivers on.

Frequently you will have one or two rounds during qualifiers where your opponents are not competitive, or perhaps their robots do not even show up for the round or do not function at all when they get there. These are prime opportunities to run up your ranking points. You should not sit back and relax during these "gift" rounds. Instead, you should score as many points as you possibly can early in the game, then assess carefully how many points you should score for your opponents in order to gain a solid number of tie breaker points.

Obviously there is some judgment required here. You do not want to score for your opponents unless you are certain you are way, way ahead and can afford it. You also need to consider the rules of the specific game. For example, in the 2009/10 season there was a special "doubler ball" which complicated the risks of scoring for your opponent, since you could not be sure if the opponent would successfully double the points in their goal at the very end of the game. In this case, you had to be certain that even if the opponent doubled a goal they still would not catch you.

To show you how important it is to pay attention to ranking points, consider the 2009/10 New York City Championships. Going into the final round our team and one other team were undefeated. The other undefeated team needed 40 ranking points in their final game to be ranked first (ahead of us) based on tie breaks. One of their opponents did not even show up at the match, giving them a two on one advantage, and their other opponent was a "low goal only" design that was not competitive. But the undefeated team did not seem to realize what the tie break situation was. They ran up a huge score over their opponents (something like 100 to 3) and did not toss in a single opposing goal to gain ranking points. As a result, our team was the first seed captain team and they were second. This gave us the crucial first pick of alliance partners, and ultimately led to our victory in the championship. Had they been aware of the situation, they probably had enough time to score the 40 ranking points required to become first seed (i.e. by winning 60 to 43 rather than 100 to 3) and the outcome may have been far different.

 # 374. STRATEGY: Communication with your partner team is crucial for success.

Your partner robots are randomly selected during qualifier rounds, and so you are likely to play with partners you have never even met before! This makes communication very important, and very difficult. But understanding each other's strategy and being able to communicate effectively during the round can make the difference between winning and losing a close match.

To facilitate communication, talk to each other before the round starts (there are often five to ten minutes available while you are queued for your match). Talk about the offensive and defensive strategy, which team is going to start with any initial game elements (many games start each alliance with some number of scoring elements for use during autonomous, for example). Be sure you know whether their robot is having any hardware or software issues and what its capabilities are so you can plot your strategy accordingly. We have even had cases in which we were able to help our partner team solve a hardware or software issue while queued up for a match!

 # 375. TIP: If you need a part or tool you failed to bring, ask other teams for help. Likewise, if you hear of another team needing something you have, give it to them.

This is a part of "Gracious Professionalism." Even though you are competing against other teams, they will help you out when you are in need. This also applies to getting help to debug software or sensor issue, or any other need you may have. On the flip side, you should be very willing to do the same for others, even if that puts you at a disadvantage when you take the field against the opponent you helped. Winning on the field is always considered secondary to exhibiting "Gracious Professionalism."

 # 376. STRATEGY: Never give up, never stop, and keep fighting until the last second.

We've seen teams shoot all their game elements then just sit and do nothing for the final few seconds of a match. This can be a big mistake. Instead of just sitting there, go try to harvest one last scoring element and take a final pot shot at the goal! Or, if that's not possible, go try to prevent the opponent from making their final shot (within the game rules of course.) The point is, some games are won or lost by a handful of points, don't stop fighting until the final buzzer.

Alliance Selection

The alliance selection portion of the tournament is crucial for every team. If you have scored enough wins and ranking points to be a captain team, congratulations! You now must use the alliance selection process to choose partner teams that will maximize your chances to walk away with the top prize. If you are not a captain team, then how you handle yourself before alliance selection means the difference between making it into the elimination rounds and watching the rest of the tournament from the bleachers. Even if your record only puts you in the middle of the pack, you may still have a chance of getting into a good alliance and having one final shot at victory.

 # 377. TIP: Confidence and relationships are key.

The trick to having more teams notice you is having the confidence to stand out and speak to everyone. To be noticed, one has to make an effort, starting at the very beginning of the tournament, to form relationships with others. Robotics is not just about hardware and software, it is also about leadership, public speaking skills, and forming friendships with other teams. Put those skills to use! Don't be afraid to walk up to people in the stands and introduce your team, or strike up a conversation with the team next to you in line at hardware inspections. Getting noticed leads to relationships that could be invaluable to your success.

Think of it this way: if you were a captain team and someone you'd never spoken with before came up and said, "Please pick me as your alliance partner!" wouldn't you be less likely to pick them than someone pleasant you had been talking with throughout the day? (Assuming both teams seemed about equal in capability of course.)

 # 378. TIP: When you select an alliance partner as a captain team, or when you accept an invitation from a captain team as a partner, use traditional FIRST alliance selection phrases.

For example, as a captain team selecting a partner, the traditional wording is along the lines of: "We would like to invite team 9999 to be our alliance partner." The traditional acceptance wording would be something like: "Team 9999 graciously accepts!"

 # 379. TIP: Create a one-page "flyer" that advertises your robot's capabilities in order to "shop" your robot to the captain teams.

If you are not a captain team this can be useful when trying to convince another team to pick you as an alliance partner.

Of course, you must create this before the tournament even begins, in case you are not a captain team. Be sure to bring a number of copies, and start pitching your robot even before qualifiers are done.

The photo shown here is the flyer that one of our teams used at the World Championship in Atlanta (Fatal Error is the name of the team). The concise list of robot features is useful when captain teams are looking for robots that compliment their strategy. The bullet list of past accomplishments was helpful in convincing teams that the claims about the robot were not exaggerated. This team was selected as an alliance partner even though there were several teams with better overall records during qualifiers.

 # 380. STRATEGY: Consult your scouting reports carefully.

Don't throw away those scouting reports when qualifiers are finished, they are crucial to help you locate robots that would make good partners.

 # 381. STRATEGY: As a captain team, do not depend on the final rankings after qualifiers as your only source of information for which teams to select as a partner.

This is a mistake that we have seen many teams make. Instead of doing careful research and scouting during the qualifier rounds, they just look at the final rankings of teams after the qualifier rounds and assume that the order that teams fall in those rankings equates to how good each team would be as a partner. But the final rankings after qualifiers are frequently not a reliable way of picking the best robot for your situation.

There can be cases where one of the best teams simply has had hardware problems in their first few rounds of play, then finally fixed their robot only near the end of qualifier rounds. Their record may look dismal but their late fixes may in fact make them one of the best teams to pick. In addition, many teams' records will differ only by ranking points (tie breaks). Is a team that is seeded in sixth place with a record of 3 wins and 1 loss and 100 ranking points really that much better than a team in tenth place with the same 3 wins and 1 loss but only 50 ranking points? Maybe they just don't understand ranking points, which is irrelevant during elimination rounds, or maybe the way they were paired with partners just randomly made it harder for them to score ranking points.

The point here is, you should be scouting every robot from the start of the tournament, not waiting until 15 minutes before alliance selection to figure out which robots will make your alliance most able to survive elimination rounds. The scores teams achieved during qualifiers may be quite misleading in some cases, you need hard facts.

 # 382. STRATEGY: As a captain team, find a complimentary robot.

For example, if your robot is great at scoring but not so great at autonomous mode, find a partner who can compensate for your weaknesses. This may not always be the "best" robot remaining that has not yet been chosen.

 # 383. STRATEGY: As a captain team, give weight to robots you have practiced with.

Coordination of strategy on the field can sometimes outweigh technical capabilities of a robot. If there is a team in your area that you have practiced with and you feel they can coordinate with your team effectively, they may be a better choice than a team with a technically stronger robot with whom you have never practiced.

 # 384. STRATEGY: As a captain team, after identifying some candidate partners, go interview them. Make sure nothing has gone wrong with their robots and they are still in good working order.

We have seen cases where a robot did great during qualifiers, but perhaps had a terrible hardware failure right at the end, a problem that cannot be fixed in time for elimination rounds. The only way to find this out is to ask each candidate alliance partner if their robot is still in good shape. We did see one case where a very honest team actually declined an alliance captain's invitation, citing that their robot was no longer functioning and it would not be fair to the captain team for them to accept. But rather than rely on that, it is better to ask before picking a team.

 # 385. TIP: As a captain team, after doing all your research and interviews, prepare a list of candidate partner teams, ranked from high to low.

Don't trust your memory. When you're standing up there and the tournament announcer comes and asks which team you wish to pick, it's nice to have a piece of paper with all the team numbers neatly ranked starting with your most desirable partner team. We have seen cases where a team accidentally chose the wrong partner due to a lapse in memory; they simply said the wrong team number by accident.

 # 386. TIP: As a captain team, make sure you have enough picks on your partner candidate list.

If you need to pick two partner teams, don't just have two choices on the candidate list you bring to alliance selection, because another captain team may choose one of them. If there are four captain teams, and you are the number one seed, and you need to select two partners, you will require at least five selections on your candidate list. For example, you will get first pick, and let's assume the team you pick accepts your offer. But then the second, third, and fourth seed captains each get a pick, and they could very easily choose the second, third, and forth candidate teams on your list. So, you need a fifth team in mind in case that happens.

It gets worse if you are not the first seed. The fourth seeded captain team needs at least eight candidate partners on their list if two partners are to be chosen.

In fact, because sometimes a team will not accept your offer, it's a good idea to have at least one more selection than the minimums discussed above. This is especially true if you plan to choose another captain team (more on that in another tip below).

The following chart shows you how many teams you should have on your candidate list if you are a captain team.

Captain Team's Seed Number	Selecting One Partner (division under 20 teams)	Selecting Two Partners (20 or more teams in div.)
1	1 minimum, 2 recommended	5 minimum, 6 recommended
2	2 minimum, 3 recommended	6 minimum, 7 recommended
3	3 minimum, 4 recommended	7 minimum, 8 recommended
4	4 minimum, 5 recommended	8 minimum, 9 recommended

 # 387. STRATEGY: Pick another captain team to eliminate a stronger opponent and form a tough alliance.

In most tournaments it is perfectly acceptable and even most sensible for one alliance captain to pick another in the first round of picks. (There is an exception for some scrimmages that do not allow captains to pick other captains.)

This is not only acceptable, it is probably the best strategy in most cases. For example, if you are the number one seed, and your scouting confirms that the number two seed really is the best remaining robot, why not form an alliance with that team and create an unbeatable combination? You not only get them on your team, you no longer have to go up against your toughest competitor!

Before doing this, however, it is in your best interest, and in the interest of Gracious Professionalism, to ask that team if they will accept or not. For a championship tournament, only the winning alliance captain team is automatically invited to the World Championship. The first picked alliance partner enters a lottery and may or may not get an invitation. In the example above, if the second seeded team thinks they have a reasonable chance of defeating you (and therefore earn an automatic World Championship invitation) they may well decline your invitation. You do not want to embarrass them by making them decline your offer in front of everyone. Confirm before the selection process whether or not they will accept your offer.

 # 388. TIP: Be honest with teams who ask you to pick them as a partner but whom you do not expect to pick.

If you are a captain team you may find several teams coming up to you before alliance selection asking to be picked. You should not lead them astray. Be honest but sensitive to their feelings if you do not intend to pick them. Simply tell them that while you respect their robot, it is not the most complimentary partner choice for your strategy. If they are on your list somewhere but not in your top few choices, again simply be honest and tell them that you may pick them but only if some of your other choices are not available. You don't want to lead someone to believe you will pick them only to disappoint them later, that is not a part of "Gracious Professionalism."

On the other hand, if you are the one asking a captain team to pick you, you should try to logically explain why you believe you are the best choice for their team rather than just shouting "pick me!" If they do not end up picking you, you should not be angry about it, but rather just graciously accept their decision. It is their choice as captain team.

Elimination Rounds

If you are lucky enough to make it to elimination rounds, now the real fun starts! Crowds will usually be much larger and more boisterous, and the pressure is on. You may be exhausted from all the work it took to survive the qualifier rounds, but it's not time to rest, you will now face the toughest competition in the tournament for a series of best two out of three matches, with very little time between each game. So what do you do? The basic answer is: exactly what you have been doing, only more so!

 # 389. TIP: During elimination rounds, keep spare batteries right on your cart.

You may not have time to get back and forth to the pit. Make sure you have a system to identify which batteries are already used and which are fresh (see Electrical System, Battery Tips for methods of doing this.) Even so, you may need a couple of members of your team who are not on the driver team to keep shuttling used batteries back and forth between the game field and the pits. Keep them charging, you never know how many games you will need to play. We have even seen cases of a tie during an elimination match that made it go four games instead of the usual maximum of three, so battery management is critical.

 # 390. TIP: If you are captain team, use your strongest alliance partner first in each new level of elimination rounds.

It is a psychological advantage to win the first contest, so put your best foot forward in the first of the games against each new opponent. If you beat them in that first game they will be under tremendous pressure in game two and that may cause mistakes that you can take advantage of.

 # 391. TIP: Be prepared to face counter-autonomous defensive modes during elimination rounds, even if you did not see any during qualifier rounds.

You will probably face much better competition during elimination rounds than you did during qualifiers. Plus, good teams may have had time to program new defensive or counter-defensive autonomous modes during the competition. Be ready for anything!

 # 392. TRICK: Sometimes it is advantageous for you, even as captain team, to sit out and let your two partner robots handle a round.

This is especially true if you have one partner whose robot is very competitive or two partner robots that are extremely complimentary. This way, you can keep the competitive partner robot in every round of play. This may also be necessary if the captain team's robot has a hardware problem that will take longer than the time allotted for a repair time out to fix.

Part X: Community Outreach and Fund Raising

This section is all about how your team interacts with your community, sponsors, and the outside world in general.

FIRST stresses community involvement in several ways, including recognition via awards such as the Inspire and Connect awards. The Inspire award is one way you can qualify for the World Championship event, showing how much emphasis FIRST places on these activities.

Outreach events include activities to encourage students and mentors in your area to join existing robotics teams or to start new teams, as well as raising awareness of robotics in education in general. While it may seem counter-intuitive that you would try to create new competitors for your team in your area, you have to look at the bigger picture, and that is that FIRST is not really about the competition, it's about inspiring students. The more teams there are, the more students there are who might be inspired!

Interacting with sponsors and potential sponsors is necessary for fund raising, and fund raising is how most teams pay the bills. Funding is necessary for travel expenses, registration fees, tools, materials, spare parts, field elements, team uniforms … you get the idea. While FTC is not a terribly expensive sport on a per-student basis, having sufficient funds lets you concentrate on the robot, have all the parts and tools you need, and afford to travel to more competitions.

Outreach Events

 # 393. TIP: Contact your regional FIRST organization and volunteer to support their activities.

For example, we work very closely with NJ FIRST to support various events and demonstrations they hold throughout the year, including demonstrating our robots at the 100th Anniversary Boy Scout Jamboree, demonstrations at the Liberty Science Center during their Robotics Day, and other activities. This requires bringing the robots to the location of the event along with recruiting enough students to help the event organizers keep things running smoothly.

394. TIP: *Set up a booth at local festivals, flea markets, and other similar community events to promote robotics.*

Many communities host annual events where local businesses can set up tables or booths to show their wares. Often the cost of renting space at these events is low, and may be lower for nonprofit organizations such as your robotics team. If there is space, you may run your robot (or sparring robot) and let members of the community drive, to promote interest in robotics. You may recruit new team members and mentors, or motivate people to join teams in their area or even start new teams.

395. TIP: *Organize a miniature "tournament" for the general public, allowing spontaneous teams to form and compete.*

You will need at least two robots for this kind of interactive outreach event, more than two is better in case of mechanical problems during the day, so you may need to join forces with other teams in your area. You can either play the real season game, or create a simplified version that is easier to train new recruits how to play, or uses more limited field elements. Attendees who wish to drive the robots need a brief orientation session, where besides teaching them the control layout and game rules you must stress that the goal is to score points and not ram the robots together (otherwise your robots will be in tatters by the end of the day). This kind of event is best held post-season when you no longer are as concerned about damage to the robots.

396. TIP: *At outreach events, be very careful about allowing members of the public to drive your robot.*

This is especially true when the event is still mid-season. It is easy for a novice driver to damage the robot. Many members of the public have also seen "Battle Bots" type TV shows and they think the purpose of the robot is to crash. So, either don't use your competition robot at all for such events until the season is over (perhaps use your sparring partner robot if you have one), or at the very least keep close watch on any demonstrations where untrained drivers can take the controls. Remind drivers who are being too rough that this is not "Battle Bots" but rather a game that involves scoring goals.

397. TIP: *Create a flyer for outreach events.*

The flyer can use information from the FIRST web site about the educational value of robotics, along with information about your own team and other local area teams. Some pictures of robots in competition make it more interesting. The illustration here shows a flyer we used for a post-season outreach event. It is important in any such flyer to make the *who, what, where, when* information large and obvious, put the fine details smaller and later in the flyer.

Post the flyer about two to three weeks before the event on community bulletin boards located in food stores, post offices, places of worship, restaurants that teenagers frequent, and other places where you see community events flyers posted. Be sure to ask permission before taping up posters outside of small businesses! Go to all such places within a certain distance of the event location.

New Jersey FIRST Tech Challenge
Invites you to the
Pope John High School

ROBOTICS CHALLENGE & EXPO

NO PRIOR ROBOTICS EXPERIENCE NECESSARY!
ALL AREA STUDENTS INVITED, NOT JUST FOR POPE JOHN!

FREE ADMISSION

WHEN: ***Saturday, June 26, 2010***
9 am to 3 pm
WHERE: Pope John High School, 28 Andover Rd, Sparta, NJ
WHO: Prospective Robotics Coaches and Students entering Grades 7 - 12

Come and see what FIRST Tech Challenge is all about !
LEARN HOW TO START A TEAM OR JOIN AN AREA TEAM
YOU CAN DRIVE A COMPETITION ROBOT!
Introduction and orientation for students who wish to drive a champion robot in competition begins at 9:00am, or just come and watch throughout the day!

Come as a team, or on your own - we provide the robots

For more information contact Steve Pendergrast: pendergrast@optonline.net

For Inspiration and Recognition of Science and Technology

 # 398. TIP: Create a press release for an outreach event and distribute it to local newspapers.

A press release can help get your event picked up by local newspapers, many of whom love to run stories about local activities. Some also have community calendars that their readers use to look for family activities. The purpose of the press release is to hand the newspaper a ready-made story that you hope will run before your event and stimulate people to come check it out. Some papers may run your press release virtually unchanged, others may take facts from your release and create a story themselves. You may also want to submit a photograph that is relevant to your event, such as a picture of the season's robot. Make sure it is high resolution and looks good. This makes it easy for the newspaper to create a nice article.

Press releases normally begin with the words FOR IMMEDIATE RELEASE, then have contact information in case the reporters have questions. There is a head-ing, usually bold and in "title caps" (i.e. each word capitalized but not in all uppercase). The title should grab attention and entice people to read the release. After the title is the body of the release. The body should be written in clear, concise prose and should answer the basic newspaper story questions: who, what, when, where, why, how. The most important information should appear first in summary form, with additional details added with each paragraph. Background information on the sponsoring organization and FIRST would usually appear last.

The body traditionally starts with an indication of the location involved, usually a city and state in uppercase letters, and a date.

FOR IMMEDIATE RELEASE

Contact: Steve Pendergrast, [illegible] 973-[illegible]

Robotics Challenge & Expo Set To Inspire Careers in Technology

SPARTA, NJ. On Saturday, June 26th, Pope John High School in Sparta, NJ will host a Robotics Challenge & Expo for the general public from 9am to 3pm. Admission is free, and the event will feature robotic games open to students from any school. In these competitions, boys and girls will actually drive champion level competition robots and play a defined game that involves "harvesting" balls off a field and shooting them through various goals, among other challenges.

There will also be several 15-minute discussion sessions throughout the day covering topics such as how to join or start a robotics team, competition robot design, how robotics competitions work, robotics as an important part of education in science, and other related subjects. Adults interested in robotics are also welcome to participate as spectators and as discussion participants.

"The purpose of this event is to demonstrate competition robots in action, and to motivate students to join a robotics team or even start teams of their own," commented Steve Pendergrast, coach of two of Pope John High School's robotics teams. Pope John's robotics program is open to any area student, whether in public or private school, and Pope John is also active in assisting area schools to start their own teams.

Students who wish to compete in the June 26 event need to arrive in the morning between 9am and 10am to be briefly trained on game rules and robot controls, but spectators may arrive at any time throughout the day to watch the action and learn about robotics. Suggested ages for competition are 7th grade through high school, although some of the competitions are appropriate for younger participants. Spectators of all ages are invited to watch the competition and participate in discussion sessions.

Two of the robots that will be used in the event are major regional championship winners, with one robot, "Prudence," winning the New Jersey State Championship this past January and the other, "Penelope," winning the New York City Championship held in the Javits Center this past March. Both robots were invited to compete at the World Championships held in the Georgia Dome this past April, where "Prudence" made it as far as the division semi-finals against the strongest competition robots in the world. The World Championship was broadcast nationwide on the NASA channel.

The Robotics Challenge & Expo is hosted by Pope John XXIII Regional High School in conjunction with New Jersey FIRST. FIRST is an organization that runs a worldwide robotics league with programs spanning from early grade school through high school and involving over 15,000 teams and over a hundred thousand students in 2010. FIRST was founded by Dean Kamen, inventor of the "Segway" personal transportation device as well as medical equipment and many other inventions. The purpose of FIRST is to inspire students to pursue careers in science and technology. Over $12 million in scholarships were awarded this year for FIRST robotics participation.

####

Press releases traditionally end with several hash marks: #### so there is no confusion about whether any pages were lost in transmission.

The photo here shows an actual press release we used for an event. You can find more information on how to write a press release by doing online searches on terms like "press release style" or "press release basics."

Send the press release out to papers about two to three weeks before the event. Sending it too early means the timing may not be good to generate traffic to your event, sending it too late might mean you miss publication deadlines.

Some papers have ways to submit press releases online or by email, others still deal in paper via FAX or letter. Most these days have a web site where you can find out how they want information to be submitted.

 # 399. TIP: Don't schedule outreach events just before competitions or just after them.

Before competitions, you need time for practice and last minute hardware adjustments. After them, you will be exhausted, and your robot may need repairs!

 # 400. TIP: During outreach events, have short segments where team members describe the robot and what robotics means to them.

Students frequently give short talks at events like this, describing how the robot works, what you had to go through to build it, what happens at an FTC tournament, and how the student feels about being on a robotics team and what they have learned from it. This is a great way to practice your speaking skills, useful during judging, and also shows people how much fun robotics can be. Remember the goal is to motivate students and parents to become involved in robotics activities, either by joining existing teams in your area or forming new teams at their school, via scouting or other organizations.

But keep such speeches short and to the point; most people are coming to the event to watch the robots in action!

 # 401. TIP: Organize a demonstration event for schools, youth groups, children's wards at hospitals, or other locations.

Besides holding a fully fledged event, another way to reach your community is through demonstrations of your robot for specific groups of people. This requires a few students, an adult mentor to supervise, the robot in good working order (with fully charged batteries), and perhaps the game scoring element and some kind of goal. Demonstrations like this usually last an hour or so, and would feature a run-through of how the robot works, a brief description of the season game, and student's describing how they've done so far in the season. Then turn on the robot and run it through its paces, and watch everyone's eyes light up! Members of the general public are usually quite interested in robots and just calling up and volunteering to give a demonstration is usually all it takes to get an enthusiastic reply. Before you do this for the first time you may want to hold a practice session so everyone knows what role to play and to work out logistical details. Be sure to bring your repair kit and extra batteries.

 # 402. TIP: Run a summer camp for both your incoming rookie team members as well as the general public.

This can be both a fundraiser (assuming you charge a fee to attend the camp) as well as a way to recruit new team members and get them ready for the upcoming new season! Of course you need a place to hold such a camp as well as access to enough building materials to allow camp members to build a robot that has reasonable capabilities.

We ran a camp in the summer of 2011 to prepare our incoming freshmen members as well as some members of the general public to get up to speed on FTC robotics. The camp members learned how to build competition robots in the first few days of the camp using a somewhat simplified version of the 2010/11 season game, then on the final day of the camp they held a miniature scrimmage to learn a little about game strategy and how tournaments work.

 # 403. TIP: Bring one or more cameras to outreach events and take plenty of photos, you can use them later in your engineering notebook to document the event.

You want to be able to document your community outreach activities in your engineering notebook. For events that occur post-season, hold on to those photos for the next season's notebook. The photos can also be used future versions of your flyers for other outreach events. Be sure to take shots of participants watching demonstrations or controlling the robots themselves.

Fund Raising

 # 404. TIP: Know how much you need, create a budget.

A sponsor may ask you how much you need, and you'd better be able to give a coherent answer! A new team starting up might need several thousand dollars depending on what resources you already have. For example, if your school has a metal shop that already has great tools; you won't need those things in your budget. Some schools provide bussing for school activities and so you may or may not need to have travel expenses included in your spending plans.

You need to purchase the competition kit or yearly season upgrades, cover your FIRST team registration fee, tournament entry fees, materials to build a field, team t-shirts or other uniform elements, pit decorations (which can be minimal), and other items.

From our experience, you can run a competitive team that goes to several scrimmages and tournaments for about $3,500. That's about $350 per student, which is not an enormous sum considering the amazing amount of knowledge students gain from FTC robotics. In fact, it's a bargain! That's probably less than the football team spends on mowing the field for a season.

If you are a returning team and you do not need to buy a full competition kit, and if you reduce the number of tournaments you participate in, and you don't build a field, the amount would be greatly reduced, but probably not less than about $1,500 for a season.

 # 405. TIP: Consider charging a nominal activity fee to students to cover part of the expenses.

In the 2010/11 season we charged a $65 activity fee per student. This is not a daunting amount for most people and covers at least the yearly FIRST registration fee, team t-shirts, and some spare parts. We have heard of some teams charging higher fees, and of course some charge no fee to students, but this is something you should consider to get the season off the ground even before you've had time to raise money from sponsors.

 # 406. TIP: Prepare a presentation, ten minutes or so, that emphasizes the educational value of robotics.

Business people respond to value propositions. Prepare a presentation that talks about America's competitiveness in the world and how that depends on technology education. Then use material you can find on the FIRST web site to show how robotics is a key learning experience for students in technology. If you have an existing team and can demonstrate a robot, work that into the presentation as well to show what you are doing. Maybe even allow someone in the sponsor's company to drive a little! Finally, ask for a donation to support your team. It's more effective if students do most or all of the talking, so practice your presentation.

 # 407. TIP: Prepare a one page flyer that re-iterates the points made in the presentation.

This can be a hand-out that you give to the potential sponsor during the presentation, and can also be used to solicit funds from sponsors who do not have time for a full presentation.

 # 408. TIP: Call local businesses, and offer to come in to discuss the robotics team.

Many businesses love to interact with the community by sponsoring teams of all kinds. They may not be as familiar with robotics as they are with little league baseball or other sports, so offer to come in and tell them about it! Many people are curious about robotics and you may be surprised how many say "yes!" to such a presentation.

 # 409. TIP: Write a thank-you letter to every sponsor, no matter how small.

And not just an email, a real letter! You should send it as soon as possible after the donation is made, and it should be signed by all the team members.

It's also nice to send a note at the end of the season to let the sponsors know how you did and to thank them once again for their support. Even if you didn't have a great season based on scoring wins, emphasize how much you learned and how you have plans to improve next season. If you had a great season, of course, a photo of the team holding a trophy or two with the robot would be a nice touch!

 # 410. TIP: Bring your robot to fundraising presentations for potential sponsors.

If you have a working robot be sure to bring it and demonstrate it. We let one sponsor drive our robot in his conference room, and he had a grin from ear to ear as he took the controls (plus he later made a generous donation!). Of course if you are just starting a rookie team, you may not have a working robot when you make your first presentations. But if you have one, bring it.

 # 411. TIP: Put sponsor logos on your robot, on your pit banner, and other places.

This is usually reserved for the larger donations since there will be limited space, especially on the robot itself, but this is a great way to show your appreciation to the major sponsors. If you take photos of the robot to send to sponsors, of course make sure their logo is showing in the picture!

 # 412. TIP: Several small fundraising events are easier to coordinate than a single large one.

If you organize a specific event to raise money, keep in mind that the difficulty in organizing an event goes up exponentially with its size. You also may be putting "all your eggs in one basket" if you try to run one mega-event. Several smaller events might be simpler to coordinate and result in better results.

 # 413. TIP: Invite potential sponsors to come visit your school (or other organization) and give their own presentation about what they do.

For example, arrange for a speaker from the sponsor's company to come in and give a presentation for your school's engineering, math, science or business class sponsored by the robotics team. Drawing the sponsor into your world makes it more likely they will see the value of robotics and become a sponsor, if not in the current season then maybe the next!

Sponsorship Other Than Cash Donations

Sponsors can help your team in many ways besides cash donations. Here are some other ways you might suggest for sponsors to support your team if they are not in a position to donate cash.

 # 414. TIP: Ask sponsors about providing technical or non-technical mentors for your team.

Many companies encourage volunteerism among their employees. Don't be afraid to ask the potential sponsor to post a call for volunteers. Many engineers are just itching to work on a fun project like robotics!

 # 415. TIP: Ask sponsors to provide services such as laser-cutting, CNC machining, graphics work for robot or pit decorations, food, t-shirts, materials, office supplies, etc.

We received some extremely useful laser cutting services from one sponsor in the 2009/10 season. Many different kinds of companies have this kind of capability. Think about sign companies that might donate a beautiful set of team numbers ready to mount on your robot or a banner for pit decorations. Maybe a local copy center would donate duplication services to print flyers for your outreach events. There are many possibilities. Several of our local food establishments donate snacks for our long build sessions a few times per year.

416. TIP: *If they are still unwilling to donate anything outright, ask for a discount.*

If a business can't donate a material or service, how about asking for 25% off something the team needs to help out? If you don't ask, the answer is always "no!"

417. TIP: *If a sponsor has an appropriate space, ask for a donation of space for meetings or a competition field.*

In the 2009/10 season a sponsor donated a huge piece of warehouse space for our competition field and that made a big difference for our team.

418. TIP: *Don't be proud: ask for used computers, laptops, filing cabinets, etc.*

A used laptop several years old is probably still just fine for running the field control system and doing some programming, and even if it's not your main computer, it's always good to have a spare at the tournament in case of disaster. An old table may be junk to a small business, but it may be a fine workbench for your workshop.

419. TIP: *Ask sponsors to mention your team in the public relations or on their web site.*

Several of our sponsors were happy to post pictures we sent them of our team and robot to their web sites. This gives your team more publicity which may help you recruit other sponsors, and reminds those existing sponsors that they have done something good to help out your team.

Appendix A: Useful Web Sites

Some of the tips in this book have specified good places to buy tools, parts, materials, etc. This appendix summarizes all of these web sites for your convenience.

FIRST Robotics, FIRST Sanctioned Parts Suppliers

www.usfirst.org	Official FIRST web site. Click on the FTC link.
parts.ftcrobots.com	Purchase Tetrix parts, chain, tools, etc.
www.matrixrobotics.com	Alternative legal building kit.
www.lego.com	Purchase NXT bricks, Mindstorm sensors, Lego parts.
www.robotc.net	Robot C software download.
www.hitechnic.com	Mindstorm compatible sensors, motor controllers, multiplexors, etc.
www.ni.com	National Instruments, makers of Labview. There are usually specific FIRST versions of Labview for FTC.
www.andymark.com	FTC field rails, often game elements, pulleys, timing belts, gears, etc.

FTC Related Tutorials and Information

www.tetrixrobotics.com	Tetrix building tutorials. Click on the link for "Building System" then select a tutorial on the left side of the page.
www.hitechnic.com	Free "experimenter's kit" download on the downloads page gives lots of examples of using the ProtoBoard.
www.chiefdelphi.com	Robotics forums that many teams participate in. Several robotics leagues represented, not just FTC.
www.firstnemo.org	Non-Engineer Mentor Organization. Website has a great deal of information for non-technical mentors, but also many pieces of advice for teams.
code.google.com/p/libftc-util/	Pope John XXIII High School repository of useful RobotC library functions for FTC program-

ming.

Tools Suppliers

www.microfasteners.com	Driver handled hex keys, #6-32 taps, etc.
www.harborfreighttools.com	All kinds of tools.
www.sciplus.com	Surplus tools, other assorted items possibly useful for pit decorations and other purposes.
www.wihatools.com	Nut drivers, T-handle hex keys, 5/16" wrenches, and more.
www.elexp.com	Soldering kits, wire strippers, and other electrical tools as well as components for the Prototyping Board.

Building Materials and Fasteners

www.mcmaster.com	Almost anything you want for building!
www.microfasteners.com	#6-32 Screws, nuts, washers in bulk quantities.
www.mcmaster.com	Plastic sheet, aluminum sheet, nonstick pad, surgical tube, and other materials.
www.onlinemetals.com	Aluminum sheet, some plastic as well.
www.usplastic.com	Plastic sheet.
www.thebigbearingstore.com	#25 chain, chain breaker tool, half and full links.
www.microrax.com	Aluminum slotted extrusions and brackets.
us.misumi-ec.com	Aluminum slotted extrusions and brackets.
www.tapplastics.com	Plastic benders, tools, and adhesives.
www.delviesplastics.com	Plastic benders, tools, and adhesives.
www.amazon.com	Amazon.com has increasingly added building materials and industrial supplies. In particular, at this writing it is one of the best places to get 3D printer filament.

Electronics Components/Custom Sensors (for use with Protyping Board)

www.elexp.com	Electronix Express, switches, sensors, resistors, capacitors, LEDs etc. Color coded wire at low prices.
www.digikey.com	Incredible assortment of electronic components including many specialized sensors.
www.goldmine-elec.com	Electronic Goldmine, low cost surplus components.
www.jameco.com	Jameco, low cost components, PC boards, wire, etc.
www.radioshack.com	About the cheapest source of PC boards you can find, plus other components.

Miscellaneous

www.youtube.com	Around October to November you will be able to find videos of scrimmages or actual tournaments for the new season's game. Some teams even post early experiments of their designs. It is useful to see what other teams across the country are doing, and which kinds of mechanisms seem to be working best. You can of course also post your own experiments and tournament videos.
www.twitter.com	Using twitter during tournaments can help broadcast results to team members and parents back at home.
www.unfuddle.com	Online project management software you can use to organize your projects. There is a free level of membership that would be sufficient for a single FTC team. Free Subversion and Git source code control hosting too.
www.uline.com	Cube boxes in all sizes, buy 19" and 18" and pack one inside the other for secure air travel.

Appendix B: Bluetooth Connection Procedure

By following the procedure outlined here you will reliably connect to your NXT using Bluetooth and maximize your chances of retaining your connection during game play.

NOTE: WiFi is now the method of choice for communicating with the robot (required in tournaments) and uploading programs, but occasionally it is still more convenient to use Bluetooth, so we have left this information in this edition of the book.

Before You Start

- Make sure that inactivity sleep mode is disabled on the NXT in the settings panel.
- Always plug the Bluetooth dongle in the same USB port on your laptop. Plug it in before booting your laptop.
- Do not put the Bluetooth dongle on a USB hub, use a direct connection to one of your laptop's USB ports.
- Do not have both your programming environment and the FCS connected over Bluetooth at the same time, as the programs compete for control and results in an unstable connection.

Step by Step Procedure

1. Run the FCS. Choose to connect through Bluetooth

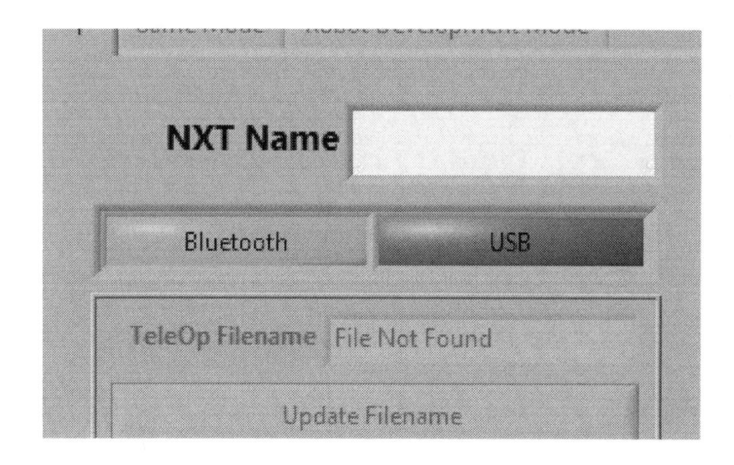

2. Enter team name.

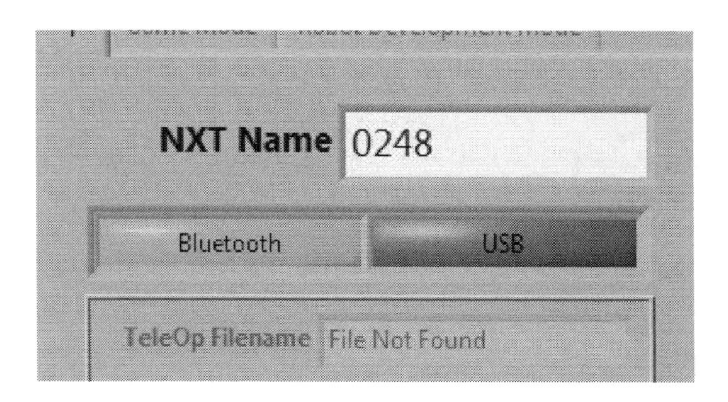

3. Make sure that Bluetooth is enabled on the NXT and press the CONNECT

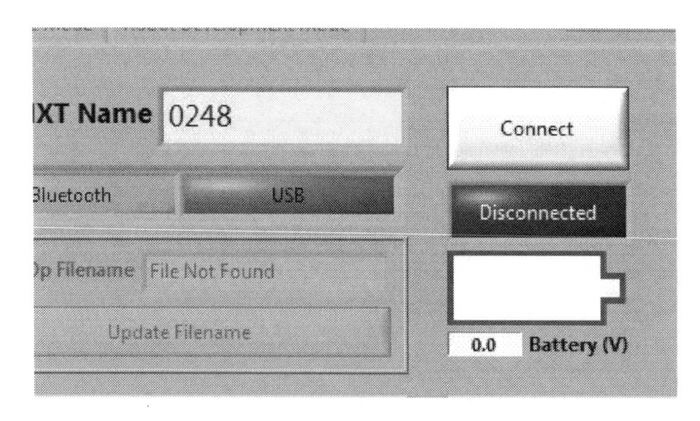

button in the FCS.

4. If a connection is established then proceed to Step 5, otherwise proceed to Step 7.

5. Press "Setup New Gamepads" and follow the onscreen instructions:

6. Choose the "Game Mode" tab, and you are done!

7. If a connection is not made to the NXT using the FCS CONNECT button, right click the Bluetooth icon in the taskbar

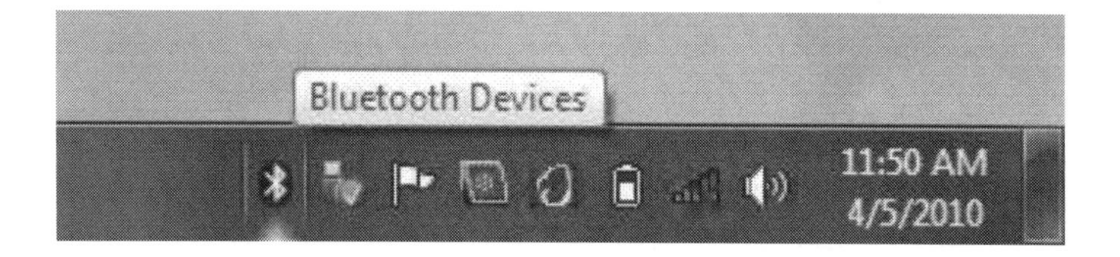

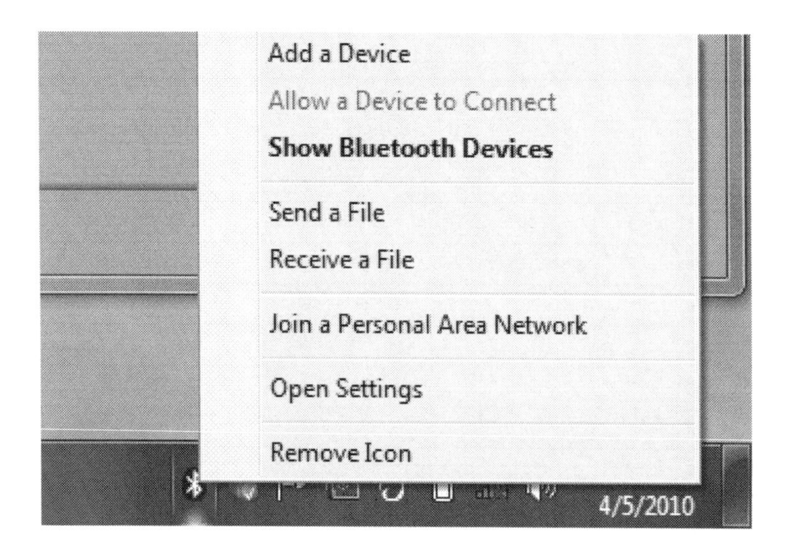

8. Select "Add a Device" and then select the NXT that you are trying to connect to.

9. When Windows prompts for a passkey, type in the passkey set for your NXT. The default is 1234.

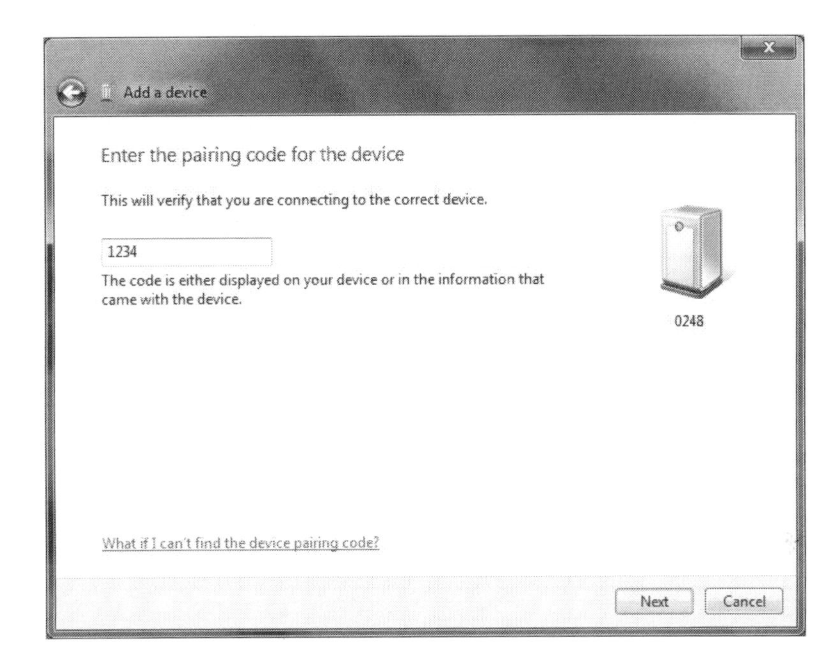

10. After you have successfully connected through Windows, try to connect in the FCS again by using Step 3.

Add a device
Select a device to add to this computer
Windows will continue to look for new devices and display them here.
0248
Bluetooth
Other
What if Windows doesn't find my device?
Next
Cancel

Appendix C: Source Code Control Systems

A useful addition to backing up your source code to external media is using a source code control system to keep track of all changes to your software over time. A source code control system allows you to compare any two versions of your code, retrieve a past version, or merge together changes made by more than one person.

There are many such systems, but two that are widely used are Subversion and GIT. These are both systems maintained by the open source community and they are free.

One way to get up to speed quickly is to use a free hosting service. These services will host your Subversion or GIT source control system on the Internet. You can upload and download copies of your source code wherever you have an Internet connection. And because your code is backed up on the web, you don't have to worry about disaster recovery if your laptop is damaged, lost, or stolen. Your code can be recovered at any time.

Some services that do free Subversion and GIT hosting are:

- Unfuddle.com
- ProjectLocker.com
- CodeSpaces.com

NOTE: we used Unfuddle.com in the 2009/10 season with a Subversion repository and we were very happy with it. We have not personally used the other services, (they seem to have good reputations on the web) but do your own investigation! There may also be other free services that are similar.

Using these services is simple. Just go to their web site, create a free account, and follow instructions for setting up your Subversion or GIT repository.

Once your repository is set up, you may perform operations like putting new software into the repository, retrieving software from the repository, uploading new versions, or comparing two different versions of a given program.

There are many different "clients" that can connect up with your online repository and allow these operations. One very popular one for Subversion is TortoiseSVN. This is a free program you can install on Windows based systems, and it integrates directly into Windows Explorer. After installation, you simply use Windows Explorer by right-clicking on your software files and using new menu items it installs to perform operations like checking a program in to your repository, retrieving a version, etc.

You can find more information at http://tortoisesvn.net

But, as we said, there are many Subversion and GIT clients. You can find a comparison of Subversion clients here:

- http://en.wikipedia.org/wiki/Comparison_of_Subversion_clients

Information on GIT can be found here:

- http://en.wikipedia.org/wiki/Git_(software)

You can also find loads of information comparing these different systems by simply searching the term "Subversion GIT Comparison" on your search engine. There are tons of tutorials showing you how to use these systems, simply search "Subversion tutorial" or "GIT tutorial" to find them.

The systems are fairly simple to use and well worth the short learning curve. These systems are used by many professional software engineers, so learning to use them on a real FTC project is a great idea.

Appendix D: Creating Sound Files for Use with NXT and RobotC

In order to play custom sound files on the NXT brick, you need to follow these steps:

1. Record the desired sound using any utility that can create WAV files. These include the windows sound recording application as well as several free and shareware utilities or commercial programs such as Sound Forge. Searching www.download.com for sound recording software is a good way to find shareware programs safely (download.com is run by CNET and they virus scan all programs they list). We use GoldWave and it works great.
2. If possible, maximize the volume of the recorded WAV file using your sound program. Trim any silence that occurs before or after your desired sound to make the file as small as possible. Convert it to be 8000 samples per second or less to save on file size. You may need to experiment to find a sample rate that gives you reasonable sound quality but is small enough to fit on the NXT.
3. Install "wav2rso," a simple utility program that can convert WAV files to the RSO format that is needed on the NXT brick. You can find this at http://bricxcc.sourceforge.net/wav2rso.zip and just unzip the file to get the executable.
4. Run wav2rso.exe and click to Select Files. Make sure the RATE is set to the same sampling rate you used when you saved the WAV file, otherwise it will play too fast or too slow on the NXT brick. You can safely also choose the COMPRESSED option, which will work on the NXT and will also result in

smaller files, saving space. When you have everything selected, click on CONVERT and the RSO files will be created.

5. Use the RobotC console to upload the sound files to the NXT. In the current version of RobotC this is done by using the menu item ROBOT, select NXT Brick, and then select File Management. You can then browse for sound files and use the UPLOAD button to load them on the NXT brick.

Appendix E:
Sample Team Handbook

TEAM HANDBOOK

Version 2a

June 1, 2011

Purpose

This Handbook is an informational guide book for the conduct of the Pope John Robotics Team members. It contains team rules, lettering requirements, and other essential information for all Pope John Robotics Team members. All team members and their parents are expected to review this material and understand it.

Pope John supports three levels of FIRST Robotics competition, resulting in a comprehensive Robotics program:

- For fifth through eighth graders in The Academy of Sussex County Catholic Schools and other area grade schools we provide space and mentor support for FIRST Lego League (FLL) as well as hosting a New Jersey Qualifier Tournament for FLL. Robots in FLL are approximately six inches cube and weigh a pound or two.
- For freshmen and upperclassmen we currently have two teams in the FIRST Tech Challenge (FTC) league. FTC robots are typically 20 to 30 pounds and fit in an 18" cube.
- For sophomore through senior high school students, Pope John has a FIRST Robotics Competition (FRC) team. FRC robots weigh over 100 pounds and fit in a cube several feet on each side. FRC is the "prestige" event in FIRST Robotics.

This handbook applies to the FTC and FRC leagues, which serve high school students.

Pope John's robotics programs are not restricted to Pope John students, any student from area schools may also participate. However, any student who participates is bound by Pope John's rules of conduct whether on Pope John grounds or participating at any event where that student represents the Pope John Robotics Team.

Goals

Our goal is to provide High School Students with the opportunity to learn from mentors who have both technical and nontechnical backgrounds, enabling them access to an unparalleled learning experience in all aspects of robot design, construction, programming, and competition. Through all our programs, we seek to instill FIRST's ideal of *Gracious Professionalism*, which holds that even when in competition every student will be courteous and even help opponents to overcome technical issues. This ideal includes participating in community outreach activities, helping to support other teams in our area, and "giving back" to our community.

To summarize, our goals are to:

- Promote and maintain a safe working environment.
- Challenge students and promote education in Science and Technology.
- Promote *Gracious Professionalism* and the Ideals of FIRST.
- Provide the opportunity to build leadership and teamwork skills.
- Use each student's unique talents to their maximum effect.
- Improve our community through outreach projects.

Team Organization

Pope John Robotics is guided by volunteer mentors who have engineering, computer science, or mechanical industry backgrounds, as well as nontechnical mentors who help organize fund raising, pit decorations, and other important auxiliary functions. In all cases the purpose of the mentors is to teach, guide, make suggestions, and organize, but it is the students who design, build, program, and drive the robots in competitions.

In addition, we have two faculty advisors, Mrs. de Vries and Mr. Ferrise, who advise the teams on both technical and administrative matters, help organize paperwork, obtain permissions and bus transportation, interface with school budgeting administrators, organize fund raising, and other important activities to keep the teams functioning smoothly in the Pope John XXIII High School ecosystem.

Season Overview

This section outlines a typical season for the Pope John Robotics Teams, although precise dates will vary and are generally not known until a season starts and FIRST begins posting tournament dates.

- FRC Season Overview:
    - *Early in the school year:* Cleanup day to organize tool shed and workshop.
    - *Late October:* Brunswick Eruption Tournament, a post-season event that uses the prior year's robot.
    - *December:* FRC team members help run the Pope John FLL Qualifying Tournament.
    - *Early January:* The new game is announced and a six week, intense build season begins. Build meetings may occur 7 days per week, although not every student is expected to attend every meeting.
    - *Late February:* The robot is shipped to a state championship event, typically either the New Jersey State Championship or the New York City Championship.
    - *March:* FRC competes at either the NJ or NYC Championship.
    - *April:* If the FRC qualifies at the championship, they attend the World Championships which will be held in St. Louis in 2011 through 2013.

  - *May:* FRC attends a post-season event, typically "Monty Madness."
  - *Late May:* FRC hosts demonstrations at "Sparta Day" festivities.
  - *Late May or June:* In recent years we have had a Robotics Team luncheon or dinner to celebrate the end of the season. Varsity letters have been awarded at this event since the 2010/11 season.

In addition to these activities, FRC will engage in various fundraising efforts during the season, support Pope John's open house, and other efforts.

- FTC Season Overview:
  - *July through August:* FTC volunteers support one or more Outreach events, typically in conjunction with NJ FIRST.
  - *Early in the school year:* Cleanup day to organize tool shed and workshop.
  - *September:* The new game is announced and the build season starts in early to mid September. Typically FTC holds two meetings per week, one immediately after school on a weekday lasting 2 hours, another on the weekend lasting 3 hours.
  - *October through December:* The FTC teams will participate in at least 2 scrimmages sometime during this time period. They may also participate in one or more "qualifier" tournaments (starting in 2010/11 both Pennsylvania and New Jersey switched to a Qualifier system).
  - *December:* FTC Team members volunteer to help run the Pope John FLL Qualifying tournament.

- *January:* FTC competes at the New Jersey State Championship.
- *February through March:* FTC will compete in at least one other championship, for example the New York City Championship, the Pennsylvania Championship if they qualified during a Pennsylvania Qualifier tournament, or perhaps the New York State Championship. The specific tournaments that are selected will depend on circumstances and scheduling.
- *April:* If one or both FTC teams qualify, they attend the World Championships which will be held in St. Louis in 2011 through 2013.
- *May:* FTC teams participate in a post-season event, usually "Monty Madness." FTC Team members who are thinking about moving to FRC for the following season will attend the FRC portion of the event.
- *Late May:* FTC volunteers support the "Sparta Day" demonstrations hosted by FRC.
- *Late May or June:* In recent years we have had a Robotics Team luncheon or dinner to celebrate the end of the season. Varsity letters have been awarded at this event since the 2010/11 season.

In addition to these activities, FTC will support fund raising efforts, support Pope John's open house and freshman orientation, and other outreach efforts.

As you can see from the information above, there is Robotics Team activity throughout the school year and even summer months. The FRC team has a short, very intense build season and will participate typically in one Championship event and two post-season events. The FTC Teams have a much longer build season spanning almost the entire school year, and will participate in many more events, typically 2 scrimmages, 2 to 3 Championship tournaments or qualifier tournaments, as well as a post-season event, and six or more outreach events. Both FTC and FRC may or may not qualify for the World Championship event in any given season.

For this reason, Robotics is a very large commitment for students to make. While it is possible to participate in a Fall sport without interfering too much with robotics activities (especially for FRC whose build season does not start until January), it would be difficult for students involved in a winter sport to fully contribute to Robotics at the same time.

Safety

The responsibility of safety lies with each and every member of Pope John Robotics. Each team member is required to abide by safety rules at all times.

It should be understood that the student members of Pope John are viewed by the mentors as young adults, acting much like employees of an organization. It is important to maintain a positive image to our community, sponsors and FIRST, and this begins with SAFETY. All Pope John Robotics Team members must know and demonstrate safe and professional behavior wherever the team conducts its business.

Important safety rules include (but are not limited to):

- SAFETY GLASSES MUST BE WORN AT ALL TIMES DURING BUILD SESSIONS, EVEN BY STUDENTS WHO ARE NOT USING TOOLS. Clip on side guards are available for students who wear glasses. Regular glasses without side guards are NOT safety glasses.

- No open toe shoes, “Crocs”, etc. Shoes provide important protection in case of dropped tools or materials.

- No loose fitting clothing, dangling jewelry, or other items that present an entanglement hazard around equipment.

- No "horsing around." Work with tools and machines does not mix with practical jokes or foolish behavior.
- No personal electronics with earphones—you need to be able to hear warnings or instructions from the mentors and other students.
- Always disable the robot's "kill switch" or remove the battery before undertaking repairs or modifications.
- Use common sense. Be aware of what is going on around you.

Robotics Workshop

Team members will primarily work in the Pope John Rectory basement, in several rooms designated for use by the robotics team. The rectory is a residence and therefore students must not wander to other areas or behave in a manner that is disruptive to others in the rectory, i.e. by making unnecessary noise. The robotics area also must be kept neat, garbage cans must be emptied, food wrappers or drink bottles cleaned up each day, etc.

Students and mentors must sign in and sign out every time they come to the rectory. Students will not work in the rectory unless at least one mentor is present. At no time will a mentor be alone with one student, there must always be a third person in line of sight.

Only mentors will have access to keys.

Tool Usage

The use of power tools is a necessity for most members of the team. Tools should always be used in a safe manner. Only adequately trained team members may use tools–*no exceptions*. Always remember that tool usage is a privilege and not a right, so use them responsibly. Talking to a mentor before using a tool is required. When using tools, a student is responsible to ensure everyone near them is wearing safety glasses and is paying attention, that there is adequate space, an uncluttered work area, and other common sense precautions.

Transportation

Transportation to and from meetings is the responsibility of each individual student. Robot Team meetings are held both after school and on weekends. (No meetings are held until Friday during the week of midterms.) The schedule is communicated via email by the coach/mentor. For FTC, most meetings are 2 to 3 hours long, but near tournament time build meetings may be longer than normal. For FRC, because of the short build season, meetings may run up to 7 days per week and may be 4 to 6 hours long or even more, but not every student is expected to attend every meeting, and homework should be brought to the meetings and can be worked on during "down time" when the student's subsystem cannot be worked on.

Students must ensure their parents arrive to pick them up on time when meetings end. Because no mentor may ever be alone with one single student, it may be necessary on some occasions for a parent who is picking up the second to last student to remain on the grounds until the last student's parent arrives. (This may happen on those occasions when only a single mentor is available for a meeting.)

Transportation by bus to competitions and scrimmages will normally be provided by Pope John, however in some cases a nominal transportation fee (on the order of $15) may be charged, depending on budget considerations.

In the event that any Pope John Robotics Team qualifies for the World Championships (to be held in St. Louis in 2011, 2012, and 2013) the cost of transportation, food, and lodging is the responsibility of the student's parents.

Academics and Scholarships

Pope John Robotics promotes strong academics; therefore it is important for students to maintain academic excellence. In all situations school comes before Robotics. Students should bring homework assignments with them to meetings (especially FRC) so that they may work on their assignments when there is "down time" caused by the need to perform certain build tasks in sequence.

FIRST actively promotes monetary scholarships from major Universities. In the 2010/11 season over $12 million in scholarships were available for FIRST robotics team participants, many from major engineering universities.

Behavior

Pope John Robotics treats each student member as a respected young adult. Each student is expected to be respectful of all team members as well as other individuals that the team comes in contact with. Professionalism is a must from each student. Students are also expected to adhere to all Pope John XXIII High School rules of conduct at all times, whether at build sessions, tournaments, outreach events, or any event where the student represents Pope John. Even students on the team who are not enrolled at Pope John XXIII High School must adhere to Pope John's rules in order to participate on the team.

Cell Phones and PDAs

Cell phone usage during meetings should only be for serious purposes related to the robotics team, such as informing a parent of a schedule change that requires an earlier or later pickup time, etc. During build sessions students should not be texting, talking on the phone socially, playing handheld games, or other activities not related to the Robotics Team. If a student has "down time" waiting for some other team to complete a task before they can proceed, they should either work on related activities (fund raising, organizing the workshop) or homework they have brought to the meeting.

Food and Drinks

Because of long hours during build season some reasonable food and drink consumption is allowable in the work area, but cleanup is the responsibility of each student and food or drinks should not be consumed near machinery, computers or other places where it may represent a hazard.

Disciplinary Actions

Violation of Pope John XXIII rules of conduct, safety rule violations, or other inappropriate activity may result in suspension or even ejection from the Pope John Robotics Team, at the discretion of the mentors and in consultation with faculty advisors on a case by case basis.

Health Information

If any student has any health issue that the mentors should be aware of it should be brought to their attention by the student's parents.

Uniform for Competition

Students will receive team shirts and are expected to wear them to all competitions, scrimmages, and outreach events, as well as team pictures for the yearbook. Other aspects of dress for tournaments should be appropriate, neat, and modest, and should adhere to workshop rules (no open toe shoes, no long dangling jewelry, etc., as this is a safety issue in the pits.)

For build meetings the student should wear appropriate, comfortable, neat clothing that does not present a work hazard or distraction.

Email List

It is important for both students and parents to be on the coach/mentor's email list to facilitate communication. The mailing list is used to inform team members of build session schedules, competitions, scrimmages, and other important information. If your email address changes you must inform the coach/mentor immediately to update the mailing list. If more than a week goes by without receiving any communications by email, it is each student's responsibility to investigate why he or she is not receiving team emails.

Financial / Fundraising

Each member of the team must pay an activity fee at the start of the season, which in 2010/11 is $65. In addition, there may be nominal charges for busing to some events. In the event any team qualifies for the World Championship, the student is responsible for travel, food, and lodging expenses in connection with the event. Finally, there may be an end of year banquet and there is a fee to attend that event.

Members of the teams will from time to time be involved in fund raising activities. This may include anything from making presentations for potential sponsors, to making phone calls, to writing thank-you letters to sponsors, or demonstrating the robots to the community at large to generate interest.

Student Involvement

The Pope John Robotics Team is a challenging activity that goes well beyond a typical high school club. There is a tremendous amount of information to learn and many skills our team members must quickly acquire. We are competing at the highest levels in state competitions, and in some recent seasons have even qualified to compete against the top teams in the world, an honor that only a small percentage of teams achieve.

For this reason, significant commitment is required from our students. Just showing up at meetings is not enough to create a successful team at this level of competition. Students must have an excellent attendance record at build meetings and competitions, but more than that, they must strive to achieve real results at each meeting, and constantly explore ways to improve the robot and the team itself in a proactive manner. Students will frequently not be told what to do; rather they are expected to figure out what to do themselves, perhaps with some guidance and teaching from the mentors.

Students who are very committed to the program may also qualify for both academic and Varsity Letter honors. Requirements for these awards are discussed further in this document.

Parent Involvement

Parents are invited to visit any build session, outreach event, or competition to observe the activities of the Pope John Robotics Team at any time. Parents are also encouraged to be on the email list the coach/mentors use to communicate with students so they are fully informed of all activities.

Due to Diocese rules, parents who wish to take a more active role in the team, which would bring them in closer contact with students besides their own, must satisfy several clearance requirements, including attending a training session and submission of information and finger prints for background checks.

All mentors have already been certified under these rules, which are the same for any club or activity at Pope John XXIII High School and are in place to protect every student. It is not difficult nor overly time consuming to meet these requirements, and if you ever want to take an active role in any team, activity, or club at Pope John we encourage you to take these steps to become certified. For more information on precisely how to become certified please contact the Pope John main office.

In addition, should any Pope John Robotics Team qualify for the World Championship or other tournament that requires an overnight trip, any parent who wishes to attend such a competition must also be certified in a manner similar to what is outlined above. Again, these rules apply to any overnight trip involving any set of Pope John students, and such certification is mandatory in order for you to attend such an overnight event, even if your child is not a Pope John student. Becoming certified in advance of such a competition will save you time later in the event one of our teams qualifies for the World Championship, as we have in several recent seasons, and you wish to accompany your child to the event.

Publicity

Part of fund raising and community involvement is the placement of news articles in papers or even national magazines from time to time (our team was featured in Scientific American Online in 2010 for example). Students may be interviewed by members of the press at events and their words and likeness may be printed as a result. FIRST requires a waiver signed by a parent or guardian giving permission for such press activity for each student. This is a requirement even to set foot in the arena. Pope John High School may also release materials to the press that include photographs of the students involved with the team.

World Championship Attendance

The highest honor a team can receive is to be invited to the FIRST World Championship event. Only a small percentage of FLL, FRC and FTC teams are invited. Teams may qualify only by winning a major regional event or championship (for example, the New Jersey State Championship or the New York City Championship) or by winning one of only a small number of major judged awards at such championship competitions.

While the Pope John Robotics Team has had both FRC and FTC teams invited in recent years, it can never be guaranteed. The only way to qualify is to be one of the very best teams in the nation, and competition is intense.

As such, in those years when we are lucky enough to attain this honor, only students who have been active, contributing members of the qualifying team (FRC or FTC) will be invited to attend the World Championship event. While involvement does not need to be as intense as that of a Varsity Letter candidate, team members who have a spotty attendance record, who failed to attend outreach events and tournaments, and otherwise have not made at least a reasonable contribution to the team's success will not be invited to accompany the team to the World Championship event, at the discretion of the mentors and in consultation with faculty advisors.

FIRST has announced that the 2011 through 2013 World Championships will be held in St. Louis at the America's Center Convention Complex including the Edward Jones Dome and other facilities, on the following dates:

- 2010/11 season: April 28-30, 2011
- 2011/12 season: April 26-28, 2012
- 2012/13 season: April 25-27, 2013

Academic Awards Recognition

Pope John XXIII High School may in some years honor Robotics Team members who achieve success at the annual Academic Award Banquet. Only students who have a solid record of attendance at build sessions, outreach events, and tournaments throughout the season will qualify for such recognition. The judgment of which students qualify for recognition at the Banquet will be made jointly by mentors and faculty advisors.

Varsity Letter Requirements

Both the FTC and FRC team members who are enrolled at Pope John may qualify for a Varsity Letter in Robotics. This section outlines the requirements to achieve this honor. It is not expected that every student, or even most students, will qualify for this honor. Only the most committed students who concretely contribute to team success are expected to qualify.

In order to earn a letter, the student must achieve all of the following requirements:

- ***Build Meeting Attendance***
  The student must have an excellent record of attendance at build

meetings, at least 70% attendance through the final Regional Championship tournament for a given season. (Some adjustment could be made for extraordinary circumstances at the coach/mentor's discretion.) For FTC members, who have a much longer season than FRC, this may mean you will need to attend 100 hours or more of build sessions over the course of the entire season. FRC members have a shorter season but it is much more intense and can similarly expect to spend many hours building and practicing to achieve this requirement.

- ***Outreach and Community Event Attendance***
  The student must volunteer for at least three outreach and community service events during the course of the season and must perform their duties at such events in a satisfactory manner. Events include: Sparta Day, NJ FIRST sponsored outreach events, demonstrating robots for potential sponsors, helping out at the Pope John hosted FLL Qualifier Tournament, demonstrations at Liberty Science Center, helping run programs to encourage other area schools to start teams, or other similar activities. There are normally six or more such opportunities available during a season for both FRC and FTC members.

- ***Full Participation at a Championship Tournament***
  The student must attend at least one Championship tournament and must execute their duties at that tournament in a satisfactory manner. Duties will vary depending on the student's roles on the team (for example, driving the robot, making repairs in the pits, scouting other teams, etc.) NOTE: the FRC season frequently involves only a single official competition, and therefore the coach/mentor has some leeway in the case of an extraordinary circumstance that prevents an otherwise outstanding letter candidate from attending the one official event. For example: in a case of illness that prevents the letter candidate from attending the only official event, the coach/mentor may allow that candidate to make up the missed event by attending additional outreach or post season events beyond the minimum required.

- ***Full Participation at a Scrimmage, Qualifier, or Post-Season Tournament***
  Scrimmages and post-season tournaments are also opportunities to gain skills that will contribute to team success. Qualifier tournaments (which occur more at the FTC level) may be required in some states before attending Championships. These activities, therefore, are crucial to a team's success even if they are not the official championship. Letter candidates should attend a minimum of one such event during the season, and they must fulfill their duties at the event in a satisfactory manner.

- ***Leadership Role Requirement***
  Simply showing up at meetings and events is not enough to letter. Varsity letter candidates must make concrete, observable, significant contributions to the team's success. Examples of ways to fulfill this requirement might include: being selected as a primary drive team member based on merit and fulfilling that role well at a major tournament, being the lead programmer who develops an outstanding autonomous mode program, being a lead builder who oversees implementation of a successful mechanism that gives the team some advantage, consistently creating outstanding engineering notebook entries (for FTC) that leads to judged awards or nominations, coming up with a fund raising idea that works well and following through to implement it, coordinating scouting activities at a major tournament that results in crucial information that aids the drive team, etc. The mentors will monitor activities of the team looking for these kinds of concrete contributions. Each candidate seeking a Varsity letter will be asked to write a description of their Leadership Role contributions for consideration by the mentors and faculty advisors. Candidates are also expected to attend periodic meetings where students discuss their contributions and season challenges.

- ***Teamwork and Gracious Professionalism Requirement***
  The letter candidate should be an outstanding team player with a positive attitude who would rather work hard to make great things happen than complain about bad luck, who would rather teach a

fellow student something new than blame them for doing something wrong, and who would rather lend a brand new battery to a competitor they're about to face in an elimination round than take a win by default. In short, the candidate strives, even under the pressure of intense competition, to always uphold the values embodied in *Gracious Professionalism*. They treat mentors, fellow teammates, referees, and competing students alike with respect at all times. They observe Pope John rules of conduct and safe work habits at all times. Mentors will observe students during build sessions, outreach events, and competitions to judge whether each letter candidate adheres to these ideals consistently. Letter candidates may be asked to write a description citing specific examples of how they have upheld the principles of teamwork and *Gracious Professionalism* for consideration by the mentors and faculty advisors.

Conclusion

Pope John XXIII High School has a comprehensive Robotics program that extends from grade school through senior year in high school. It is not restricted to Pope John students. In this program, volunteer mentors with engineering and mechanical backgrounds, as well as nontechnical mentors in supporting roles, guide and teach the students real world engineering and organizational concepts. The students then apply those principles to develop a highly sophisticated robot including mechanical systems, electrical systems, sensors and feedback systems, etc. They then develop software to control the robot in complex ways, and finally they use the robot to compete against other top teams in our region and sometimes nationally. If that were not enough, students support community service and outreach activities, maintain professional quality engineering notebooks, and learn to speak to the press, judges at competitions, and sponsors. No other high school level activity offers such a rich learning experience, and many colleges and universities recognize this by offering scholarships and highly valuing robotics experience on application resumes.

Pope John supports and recognizes this program as a key learning experience in science and technology by providing both academic and Varsity Letter recognition for Robotics Team members who are active contributors to the team. In addition, Pope John provides substantial support in terms of workshop space, budgeting, and administration of the team.

In recent years both our FRC and FTC teams have qualified for the World Championship event, in which only a small percentage of teams nationally are invited to compete.

Robotics requires a real commitment in time and energy from each student who participates, but the rewards in terms of education, scholarship opportunity, national recognition, and the thrill of intense competition will repay the students' efforts many times over.

INDEX

39878308R00146

Made in the USA
Lexington, KY
16 March 2015